Labster Virtual Lab Experiments: Basic Biology

Sarah Stauffer · Aaron Gardner ·
Dewi Ayu Kencana Ungu ·
Ainara López-Córdoba · Matthias Heim

Labster Virtual Lab Experiments:
Basic Biology

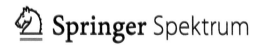

Sarah Stauffer
Labster Group ApS
København K, Denmark

Aaron Gardner
Labster Group ApS
København K, Denmark

Dewi Ayu Kencana Ungu
Labster Group ApS
København K, Denmark

Ainara López-Córdoba
Labster Group ApS
København K, Denmark

Matthias Heim
Labster Group ApS
København K, Denmark

ISBN 978-3-662-57995-4   ISBN 978-3-662-57996-1 (eBook)
https://doi.org/10.1007/978-3-662-57996-1

Library of Congress Control Number: 2018961202

Springer Spektrum
© Labster ApS under license to Springer Verlag GmbH 2018

Illustrations by: Silvia Tjong
Editor-in-Chief: Stephanie Preuß
Cover figure: © Labster ApS

This Springer Spektrum imprint is published by the registered company Springer-Verlag GmbH,
DE part of Springer Nature.
The registered company address is: Heidelberger Platz 3, 14197 Berlin, Germany

Preface

Welcome to the "Basic Biology" textbook, which is part of the "Labster Virtual Lab Experiments" series.

This book will help you to learn the key concepts of basic biology while applying your newly acquired knowledge in a virtual lab environment. In each chapter you will be introduced to one of five virtual lab simulations and the true-to-life missions that you will encounter when playing the simulations. Study the theory section presented in each of the chapters closely and you will be fully prepared to master the challenging tasks in the virtual lab!

Finally, you will find learning objectives and techniques covered by the virtual lab simulation at the end of each chapter to easily align its content with your exam preparation.

About Labster

Labster is a company dedicated to developing virtual lab simulations that are designed to stimulate students' natural curiosity and highlight the connection between science and the real world. These simulations have been shown to improve the achievement of learning outcomes among students, by making the learning experience more immersive and engaging. The content of this book was created by the Labster team members Dr. Sarah Stauffer, Dr. Aaron Gardner, Dewi Ayu Kencana Ungu, Dr. Ainara López-Córdoba, Matthias Heim, and Silvia Tjong.

About this Textbook

This book represents a great starting point for your adventures in life science. Focused on the basic unit of life, the cell, you will learn how cells divide, proliferate and provide themselves with energy and the complex compounds they need to survive. Each individual cell works in harmony with the trillions of others in our bodies, making us the incredible living, breathing, moving, thinking, and above all learning organisms that we are.

Lab Safety
In the first chapter of this book, you will learn the theory of lab safety and then put it into practice. Our Lab Safety simulation is a great place to learn about lab safety as you can experience what can go wrong in a lab in a perfectly safe environment. This way you will be fully prepared to step into other virtual and real-world labs.

Mitosis
With your newly acquired safety knowledge you will be able to move straight onto the Mitosis simulation. Mitosis is a type of cell division which occurs constantly in our bodies to create new cells whenever they're required. Controlled by the cell cycle, mitosis is typically a tightly regulated process; however, issues can arise during mitosis which lead to the development of various disorders. Will you be able to utilize the compound paclitaxel to help treat these disorders?

Meiosis
Importantly, mitosis is not the only type of cell division. The other type of cell division is meiosis which gives rise to the male and female sex gametes, the sperm and the ovum (egg). While meiosis and mitosis share some features, meiosis differs in that it results in cells with half the DNA content, this way when the sperm fertilizes the egg a normal amount of DNA is present. In the Meiosis simulation you will be able to observe meiosis, and identify all the key stages. Will you be able to apply your knowledge of meiosis to the techniques of *in vitro* fertilization and preimplantation genetic diagnosis to help a couple have a healthy baby?

Cellular Respiration
So where do our cells get the energy to divide and carry out all their other specialized functions? All cells really need to produce energy is a little bit of glucose, but give them oxygen too and their energy production increases massively. In the Cellular Respiration simulation you will learn the differences between anaerobic and aerobic respiration and how much energy each process produces. You will then

see if you can measure this in real time in a living organism using the technique known as respirometry.

Protein Synthesis
The proteins used in cell division and respiration, and thousands of other processes, are made by the cells themselves. In the Protein Synthesis simulation you will be able to follow this pathway through from the starting DNA code, which describes the protein, all the way through to the final assembled and functional protein. Along the way, you will learn how we can control this pathway to create man-made proteins to treat disease, but also how some individuals use these proteins to give themselves a competitive edge. Will you be able to combine all your knowledge and catch the doping cheat?

How to access the virtual lab simulations?

You can access the five virtual lab simulations included in this book at www.labster.com/springer.

If you have purchased a printed copy of this textbook, you will find a voucher code **on the last page**, which gives you free access to the five simulations for the duration of one semester (six months).

If you are using the e-book version, you can sign up and buy access to the simulations through the same link.

Please be aware that the six month period starts once you sign in for the first time.

If you have any questions about the use of the voucher, you can contact us at customerservice@springer.com.

Contents

Lab Safety

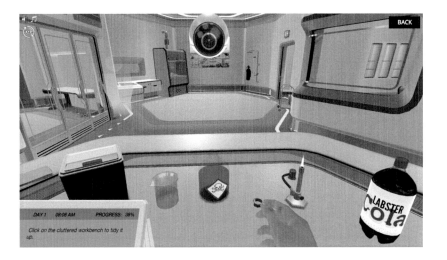

S. Stauffer et al., *Labster Virtual Lab Experiments: Basic Biology*,
https://doi.org/10.1007/978-3-662-57996-1_1

1.1   Lab Safety Simulation

Laboratories can be very dangerous, especially if you've never set foot in one before. So in this simulation, you will get the chance to make your debut in a virtual one! You will learn how to use the lab safety equipment, and how to react in case of an emergency. Start by detecting and eliminating sources of danger, then pass on your lab safety knowledge to friends.

Identify Hazards
Safety first! Always pay attention to potential hazards when you enter a lab. In this simulation, you will create a tidy and safe working environment by identifying and eliminating hazards in the lab. You will be introduced to the lab safety rules and the safety equipment, that will help you and your colleagues in case something was to go wrong in a real lab.

Emergency Training
You will be introduced to the basic hazard symbols (Fig. 1.1) used to categorize dangerous materials. You will use this knowledge to prevent dangerous situations, like acid spills. You will also learn how to deal with unlabeled, potentially haz-

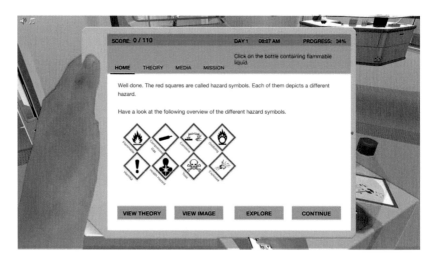

Fig. 1.1 LabPad in the Lab Safety simulation showing some common GHS hazard pictograms

Fig. 1.2 Safety equipment in the Lab Safety simulation

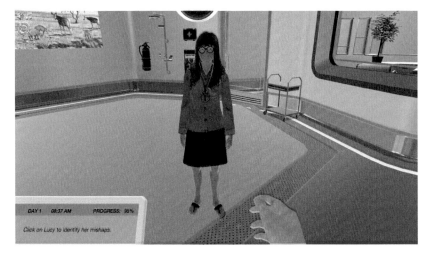

Fig. 1.3 Apply your learning to help your friend stay safe in the Lab Safety simulation

ardous chemicals. By mastering such situations in this simulation, you won't have to worry about being exposed to any real danger. You will learn how to operate the eye-wash and also get an introduction to various other pieces of lab safety equipment (Fig. 1.2).

Spread Your Knowledge
Sharing your newly acquired knowledge with your colleagues is important (Fig. 1.3). In this simulation, you will meet your friend Lucy who has never worked in a lab before. Pass on your lab safety knowledge and help her dress appropriately for a day in the lab.

Will you be able to apply your knowledge to keep yourself and your friend safe in the laboratory?

1.2 Lab Safety Theory Content

Before starting your learning adventure in the upcoming biology labs we should make sure that you have a good theoretical and practical understanding of lab safety. This way you can keep yourself, your friends and your colleagues safe in the lab, even when working with potentially dangerous reagents or pieces of equipment.

Good Lab Practices

Always remember safety comes first! Make sure you follow the safety requirements of the lab and that you are wearing the appropriate personal protective equipment. When working in a new lab, look around and identify the safety equipment so you can react quickly in case of an emergency. Always follow the guidelines below to work safely and efficiently in the lab:

- Plan your experiment in advance and ensure you have all the necessary reagents and equipment to hand.
- Keep track of all the steps that you performed and write down exactly which reagents you used.
- Make sure you label all your samples with their contents, potential hazards, the date of use and your initials.

- After you're finished with an experiment, clean all used glassware to the specification required by the lab, return reagents to the correct storage area and dispose of waste in the appropriate containers.
- Clean the workbench with 70% ethanol or another cleaning agent as specified by the lab.

Lab Safety Practices

Outside of good lab practice, there are several specific rules you should follow to ensure your safety, and that of your friends and colleagues:

- Eating, drinking, smoking, and storing food and drink in the laboratory are strictly forbidden!
- Make sure you always wear appropriate personal protective equipment for the task at hand, for example, a lab coat, gloves, and safety glasses.
- You should ensure that you wear clothes that cover your entire body. This includes closed shoes and long trousers. Avoid loose sleeves and tie back long hair as it might obstruct you and can be dangerous when working around a flame.
- Only bring the things that you need into the lab. Leave all personal items such as backpacks, purses or jackets outside, so there is no risk of them becoming contaminated. This also applies to jewelry, watches, and phones. Make sure you do not contaminate light switches and door knobs. Keep your hands clean and your nails short, and make sure you wash your hands before leaving the lab.
- The lab room should be kept clean and tidy and all emergency exits should be clear and well signed. Never place reagents on the floor. Chemicals need to be stored in appropriate cabinets to avoid accidents.
- Keep your workbench tidy and remove any items that are not needed.

Hazard Symbols

When working with chemicals in the lab, pay extra attention to the hazard symbols on the containers the chemicals are stored in. Workplace hazard symbols are easily understandable pictograms that enable you to quickly identify a hazard (Fig. 1.4). They form part of the Globally Harmonized System of Classification and Labeling of Chemicals (GHS).

Fig. 1.4 Example GHS hazard pictograms and their meaning. Hazard pictograms are designed to clearly identify the risks associated with particular substances

Safety Equipment

Labs typically contain a wide variety of safety equipment. Before beginning work you should familiarize yourself with their location in the lab, and their correct usage.

In a lab, you will typically find the following safety equipment:

- A safety shower and an eyewash station. Both should be tested weekly by a designated individual to ensure they are working properly and that the water is clean.
- A fire extinguisher, which is usually located close to the entrance.
- Two emergency exits at different sides of the lab. This ensures that nobody can get trapped in case of a fire. Emergency exits should be kept clear at all times.
- Fire blankets can be used to smother fires or protect yourself.
- The first aid kit is used for minor injuries such as cuts.
- Evacuation plans should be placed near the exits; you should familiarize yourself with these before working in the lab.
- Chemistry labs often contain fume hoods that protect the user from chemical exposure.

Lab safety equipment should be tested regularly to ensure that each item is ready in an emergency.

Personal Protective Equipment

As well as general safety equipment there are four types of personal protective equipment that you should use to protect yourself in the lab.

Lab Coat

Wearing a lab coat is compulsory in most labs. Even if you're not working with hazardous material other people in the lab may be and so you should keep it on at all times. Lab coats must be long-sleeved and buttoned up to fully protect both skin and clothes from spills. Lab coats should be worn exclusively in the lab to prevent possible contamination and should also be frequently laundered on-site or by a professional laundering service.

You should wear an additional, non-permeable, protective apron if you are working with splash hazards, volatile or reactive solutions that may easily pass through the fabric of a lab coat.

Safety Goggles

Safety goggles should be worn when working with reagents or techniques with a high splash risk. Ensure that goggles fit correctly and offer sufficient protection to the front and sides. Normal glasses are not typically considered robust enough to offer protection; however, customized prescription glasses are available. Special goggles should be worn when working with specific hazards; for example, a full face shield should be worn when working with liquid nitrogen due to the high splash risk.

Gloves

Appropriate gloves should be worn in the lab in most circumstances. Standard latex or nitrile gloves are suitable for most situations, however in some instances, for example when working with very cold reagents or strong acids, more robust gloves will be required.

Face Mask

A face mask should be worn to protect against splashes or reagents that emit fumes or dust. In this instance, it is always preferable to perform the work in a fume hood or other suitable area to prevent exposure in the first place.

Eyewash Station

Eye injuries are a serious danger when working with hazardous chemicals. Make sure you always wear appropriate eye protection when required. In case of chemical exposure, immediately go to the eyewash station and rinse your eyes for at least ten minutes. Contact lenses should be avoided in the lab because chemicals may accumulate in the space between the lens and the eye and you might not be able to take them out fast enough to avoid serious eye injuries.

Safety Shower

The safety shower should be used if you spill a large volume of a hazardous chemical on yourself or your clothes. You should get under the shower as fast as possible and pull the lever. Take off your clothes while rinsing the affected body parts.

Chemical Spill

Unlike spilling a glass of water, chemical spills have to be handled carefully. If a chemical spill occurs, it is important to remain calm, warn everybody in the vicinity and analyze the situation. It is crucial that you know what chemical was spilled and what hazards it poses before you initiate any cleanup strategy.

Depending on the amount and type of chemicals involved, spills are categorized into minor and major spills:

- **Minor spills:** Spills that can be cleaned up by lab personnel without putting themselves or others at risk.
- **Major spills:** Involve large amounts of chemicals or highly dangerous reagents. Make sure you evacuate the lab and contact the appropriate emergency personnel.

If a chemical is spilled over a person, immediately get them to the safety station and flush the affected area thoroughly. Any affected clothing should be carefully removed to limit further exposure.

Minor chemical spills can be dealt with in the lab, major spills should be dealt with by properly trained individuals so you should evacuate the lab and inform the relative safety individuals. Before you start cleaning up a chemical spill you need to identify what chemicals are involved. Depending on the chemicals different cleanup strategies need to be chosen.

In the case of acid or base spills, the spill needs to be neutralized first. Strong acids can be neutralized with baking soda (a weak base) and strong bases can be neutralized with acetic acid (a weak acid).

Corrosive Chemicals

Corrosive chemicals visibly damage or permanently change materials on contact. Such chemicals can damage materials ranging from human skin tissue to steel!

The major types of corrosive substances include strong acids, bases, and dehydrating agents.

Personal protective equipment including lab coats, gloves, closed-toe shoes and long-sleeved clothing should be worn when handling corrosive materials. Manipulation of corrosive materials should occur in a fume hood if there is a risk of exposure, explosion or chemical splashes.

Fire Emergencies

If there is a fire in the lab make sure you follow these steps:

1. Keep calm and assess the situation.
2. If the fire is small, for example a small amount of liquid in a flask has caught fire, try to extinguish it by smothering it.
3. If the fire is out of control, ensure the safety of everybody in close vicinity to the fire.
4. Raise the fire alarm and press the circuit breaker to turn off all machines in the lab.
5. If you are trained in the use of fire equipment and it is safe to do so, attempt to extinguish the fire. If the fire is out of control, or you are untrained evacuate immediately.

If a person's clothing has caught fire the most effective way to extinguish burning clothes is by rolling on the floor. Never wrap a fire blanket around a standing person because it can create a chimney effect and burn the person's face. If the safety shower is close, use it to extinguish the flames and cool the burns.

1.3   Let's Get Started

And so your virtual lab journey begins! Knowing the theory of lab safety is important, but now it's time to put it into practice in the Lab Safety simulation, before moving on with your learning. Will you be able to identify the hazards in the lab and help protect yourself, your friends and your colleagues?

Learning Objectives
At the end of this simulation, you will be able to...

- Use the correct clothing to work in the lab
- Describe the do's and don'ts in a laboratory
- Correctly use the lab safety equipment
- React appropriately in an emergency situation

ACCESS THE VIRTUAL LAB SIMULATION HERE www.labster.com/springer BY USING THE UNIQUE CODE AT THE END OF THE PRINTED BOOK. IF YOU USE THE E-BOOK YOU CAN PURCHASE ACCESS TO THE SIMULATIONS THROUGH THE SAME LINK.

Mitosis

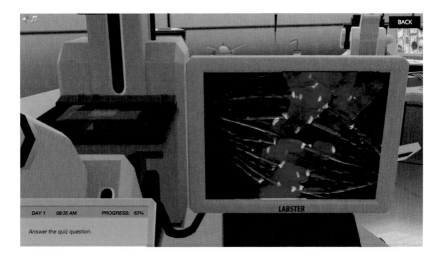

S. Stauffer et al., *Labster Virtual Lab Experiments: Basic Biology*,
https://doi.org/10.1007/978-3-662-57996-1_2

2.1   Mitosis Simulation

In the Mitosis simulation you will be able to dive into the cell cycle to understand how cell division occurs, and learn about its many different stages. You will investigate various cell types using light and fluorescence microscopy to visualize the stages of mitosis and how they are regulated. Finally, using an exciting new compound you will see if you can inhibit or accelerate cell division. How will this affect the survival of the organisms and what are the wider implications for its use?

Watch Mitosis Up Close
In the Mitosis lab, you will be able to watch 3D animations and dive into a mammalian cell to learn how eukaryotic DNA is packaged (Fig. 2.1). You will perform a series of microscope examinations, light and fluorescent, to understand the stages of the cell cycle using mammalian cells and onion root tip cells. Both mitosis and meiosis are types of cell division. In this lab, you will learn the differences between them.

Learn how the Cell Cycle Is Regulated
Learn about how the cell cycle is regulated with the help of specific proteins called cyclins and cyclin-dependent kinases. You will experiment by changing various

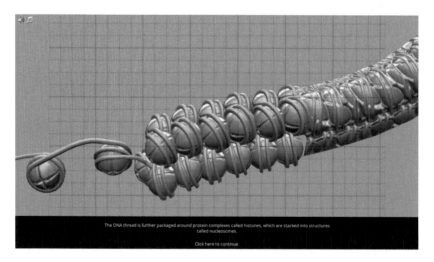

The DNA thread is further packaged around protein complexes called histones, which are stacked into structures called nucleosomes.

Click here to continue.

Fig. 2.1   3D animation of DNA packaging from the Mitosis simulation

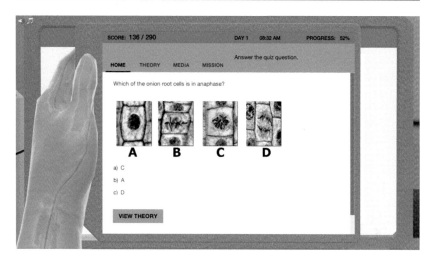

Fig. 2.2   Quiz question shown on the LabPad in the Mitosis simulation

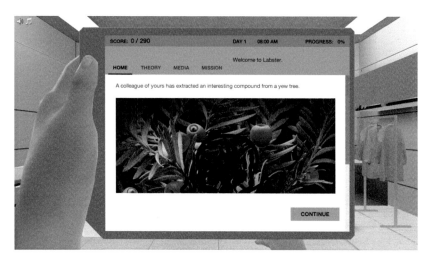

Fig. 2.3   LabPad showing the source of the novel compound used in the Mitosis simulation

cellular parameters and observe their effect on the progression of the cell cycle (Fig. 2.2). What does the cell need to do, in order to be able to divide correctly?

Test out a Novel Compound
In the last part of the lab, you will test the effect of an exciting new compound on mammalian cells (Fig. 2.3). From this experiment you will be able to determine how the compound affects cell division. Will it inhibit or accelerate cell division, how does it affect cell survival, and what are its wider implications?

2.2   Mitosis Theory Content

The concepts covered in this Mitosis theory content will cover all the topics discussed in the Mitosis simulation, and will give you a firm understanding of the basics of cell division, with later labs covering other important cell functions.

Mitosis is a type of cell division that results in two daughter cells, each having the same number and kind of chromosomes as the parent nucleus, and is typical of normal cell growth.

The Eukaryotic Cell

The cell is the basic biological unit of all known living organisms (Fig. 2.4). All cells consist of a cytoplasm contained within a cell membrane, sometimes called the plasma membrane. Beyond this, however, they can differ significantly, with major differences between prokaryotic (bacteria and archaea) and eukaryotic (plants, animals, and fungi) cells, in their structure and in the organelles they contain. Even cells within an organism can differ greatly facilitating all the different physiological functions required for life.

- **Cell membrane:** The cell membrane surrounds the cytoplasm of the cell that serves to separate and protect the cell from its environment. The membrane is composed of a double layer of phospholipids making it very flexible. It is able to host various proteins and semi or selectively permeable.
- **Cytoplasm:** The cytoplasm contains all of the material in the cell excluding the cell nucleus. Comprised of the cytosol, a gel-like substance that is enclosed by the cell membrane, and all the other organelles.
- **Nucleus:** The nucleus is one of the many organelles found within a cell. Cells typically contain one nucleus each, although certain specialized cells may con-

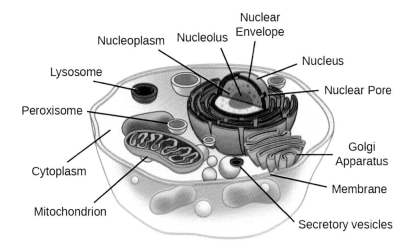

Fig. 2.4 Generic animal cell structure. Animal cells differ massively in size, appearance and function but some factors are conserved. All cells are comprised of a cytoplasm surrounded by a cell membrane. Most cells also contain a nucleus that contains a complete copy of an individual's DNA as well as other structures such as the energy-producing mitochondria and protein-producing rough endoplasmic reticulum

tain many, for example muscle cells, with others, such as red blood cells, containing none. The nucleus contains most of a cell's DNA molecules organized as multiple linear DNA molecules known as chromosomes. The nucleus is kept separate from the cytoplasm by the nuclear envelope, another double layer of phospholipids, although this membrane is punctuated by nuclear pores.

- **Mitochondria:** Mitochondria are the powerhouses of the mammalian cell generating a supply of adenosine triphosphate (ATP), for use as a source of chemical energy. The number of mitochondria present in a cell gives an idea as to how much energy it requires, for example red blood cells have none whereas cells in the liver can contain thousands.

- **Rough endoplasmic reticulum:** The rough endoplasmic reticulum is studded with protein-producing ribosomes and is the major source of protein translation in the cell.

- **Golgi apparatus:** The golgi apparatus, or golgi body is an organelle found in most cells and is a continuation of the endomembrane system and functions to package proteins for dispersal throughout the cell, or even to the outside of the cell via secretory vesicles.

- **Lysosome and peroxisome:** The lysosome and peroxisome can be thought of as the recycling centers of the cell. Both are rich in enzymes and are responsible for breaking down many kinds of biomolecules into their constituents parts for later reuse. Peroxisomes can be thought of as hazardous waste recycling centers, as a major function is to reduce the damaging reactive oxygen species into harmless waste products.

Cell Division

The only way to make a new cell is to duplicate a cell that already exists. The three main functions of cell division are:

- Reproduction
- Growth and development
- Tissue renewal

The fundamental function of the cell cycle is to accurately copy the enormous length of DNA in the chromosomes, and to segregate the copies precisely into two genetically identical daughter cells. In order to divide, a cell must first go through the following steps:

- **Signal:** The cell receives a signal for cell division related to the needs of the entire organism.
- **Replication:** The genetic material that makes up an organism is called the genome. Eukaryotic genomes consist of multiple chromosomes that are enormous in length (almost 2 m in each human cell). All of this DNA must be copied so that each daughter cell has a complete genome.
- **Segregation:** The newly replicated chromosomes come in pairs that are called sister chromatids. Mitosis is the mechanism which segregates the sister chromatids into two new nuclei.
- **Cytokinesis:** The division of cytoplasm, cytokinesis, is preceded by the division of the genetic material in the nucleus, i.e. mitosis.

DNA Packaging

Eukaryotic DNA is enormous in length, and therefore must be tightly packed in order to be manageable. DNA is wrapped around proteins known as histones,

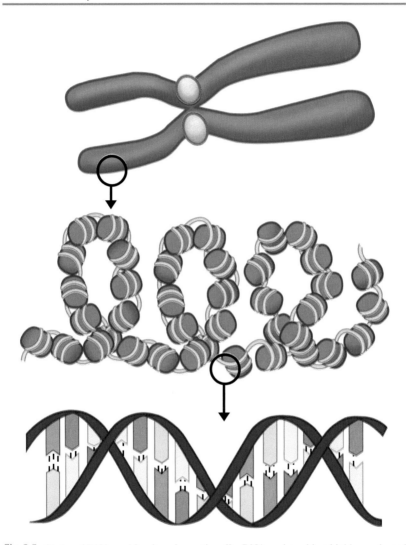

Fig. 2.5   States of DNA packing in eukaryotic cells. DNA packaged in a highly condensed chromosome (*top*), a loosely arranged string of nucleosomes (*middle*) and as a linear DNA strand (*bottom*)

forming bead-like units called nucleosomes (Fig. 2.5). Nucleosomes link together to form a string-like structure called chromatin. When cells are not dividing, the thin and long chromatin fiber is less densely packed. Prior to cell division, DNA replication takes place. After DNA replication, the chromatin is highly condensed into short and thick chromosomes forming the characteristic X chromosome shape. Each duplicated chromosome consists of two sister chromatids, attached to each other by a protein complex called cohesin that binds the two DNA strands forming a region known as the centromere. During cell division these sister chromatids separate and become classified as individual chromosomes.

Every eukaryotic species exhibits a typical number of chromosomes. Human somatic cells contain two sets of chromosomes, making 46 chromosomes in total. However, human reproductive cells, or gametes, contain only one set of 23 chromosomes.

Cell Cycle

Cells do not continuously divide. Some cells, such as nerve cells and muscle cells, do not divide at all or do so at a very low rate in adults. Cell division rates and timing are crucial for development, growth, and maintenance, of the cell and are controlled by the cell cycle.

The cell cycle refers to the series of events that occur in a cell leading to the duplication of its DNA and its division into two cells. The mitotic phase, which includes both mitosis and cytokinesis, is the shortest part of the cell cycle and for most of the time, the cells are in interphase, which is composed of three different subphases: the G1 (gap 1), S phase (synthesis), and the G2 phase (gap 2) (Fig. 2.6).

- **Interphase:** During interphase the cell can grow, develop, and function by producing proteins and cytoplasmic organelles such as the endoplasmic reticulum. DNA is replicated during the S phase only.
- **S phase:** During the S phase, the chromosomes and chromatin protein that govern various aspects of chromosomes are duplicated accurately. Chromosome duplication is triggered by the activation of a particular cyclin-dependent kinase (CDK).
- **M phase:** During the M phase, the replicated chromosomes are segregated into individual nuclei (mitosis—detailed below), and the cell splits into two (cytokinesis).

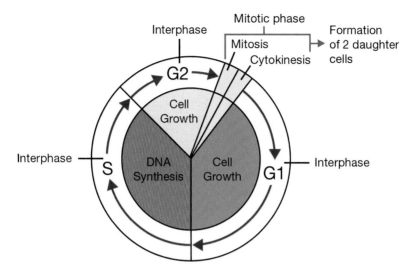

Fig. 2.6 The cell cycle. The key steps a cell must go through in order to divide. Cell division occurs during the M phase whereby the replicated chromosomes from the S phase are split in half into individual nuclei, followed by cytokinesis where the cell itself splits in half forming two daughter cells

- **Cytokinesis:** During cytokinesis, a contractile ring of two cytoskeletal proteins, actin and myosin, divide the cytoplasm into two, with each section eventually splitting to form a new cell.

Mitosis

Mitosis is the stage of the cell cycle where the sister chromatids are segregated into a pair of identical daughter nuclei, in five distinct steps (Fig. 2.7). Interphase can be considered a sixth step where the cell grows and otherwise functions normally.

- **Prophase:** During prophase, the sister chromatids are condensed. The centrosome outside of the nucleus will separate. The microtubule will form between the centrosomes (poles) to make the mitotic spindle.
- **Prometaphase:** The nuclear envelope breaks down. The spindle can now attach to the chromosomes via the kinetochore.
- **Metaphase:** The chromosomes align across the cell. The kinetochore microtubule connects sister chromatids to opposite poles.

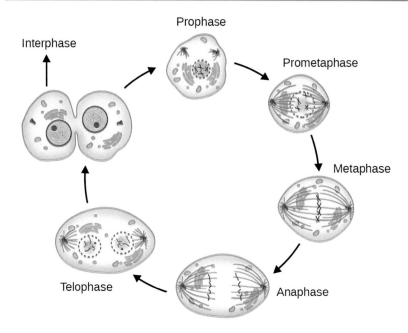

Fig. 2.7 The five steps of mitosis. Mitosis consists of five key stages, with interphase being the resting or growth stage. Briefly, during prophase the sister chromatids condense and a centrosome forms outside of the nucleus beginning to form the mitotic spindle. During prometaphase the nuclear envelope breaks down and the chromosomes can connect to the mitotic spindle. During metaphase the chromosomes align along the center of the cell with sister chromatids connected to opposite poles of the spindle. These chromatids are then separated to form distinct chromosomes as each is pulled towards the opposite pole. Finally, during telophase the chromosomes decondense and a new nuclear envelope begins to form

- **Anaphase:** The sister chromatids separate to form two daughter chromosomes and each is pulled slowly toward two opposite poles.
- **Telophase:** The two sets of daughter chromosomes arrive at the poles and decondense. A new nuclear envelope begins to form around each set.

Mitosis results in two diploid cells. Diploid refers to cells, nuclei, or organisms containing two sets of chromosomes (2n). Mitosis is one of two cell division types with the other called meiosis; which is used to generate the gametes that are haploid, containing only one set of chromosomes (n).

Cell Cycle Checkpoint

The cell cycle is controlled at three checkpoints (the G1, or restriction point, M, and the G2 checkpoint) by both external and internal signals (Fig. 2.8). These signals report whether crucial cellular processes that should have occurred by that point have been completed correctly. If all is in order, the cell may proceed to the next cell cycle phase; if not, the cell will exit the cell cycle and enter the G0, or non-dividing, phase.

- **G1 checkpoint:** This is the first checkpoint, also known as the restriction point. This checkpoint occurs in the late G1 phase when the cell begins to enter the cell cycle and chromosome duplication occurs. Cells pass the G1 checkpoint when they are stimulated by appropriate external growth factors. This prevents cells from growing in an uncontrolled fashion.

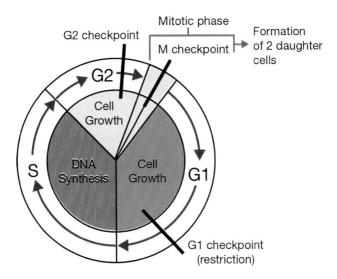

Fig. 2.8 Cell cycle checkpoints. There are three checkpoints present in the cell cycle that control its progression. The first checkpoint *G1*, or the restriction checkpoint, occurs during normal cell growth and is passed when certain growth factors are present. The *G2* checkpoint assesses for DNA damage following replication. Finally, the *M* checkpoint checks that the mitotic spindles and microtubules are properly attached to the kinetochore

- **G2 checkpoint:** This checkpoint checks for damage to DNA after replication. If any is present the cell will not be able to divide, this prevents damaged DNA spreading throughout the body.
- **M checkpoint:** The metaphase (M) checkpoint checks that the mitotic spindles/microtubules are properly attached to the kinetochore. If the cell passes this checkpoint, sister-chromatids will begin to separate, leading to the completion of mitosis and cytokinesis.

Cyclin and Cyclin-dependent Kinase

The cell cycle is mainly regulated by two types of proteins: cyclin-dependent kinase (CDK) and cyclin. CDKs are protein kinases, a type of enzyme that activates or inactivates other proteins by phosphorylating them. In order to be active, CDKs needs to form a complex with cyclin.

One type of cyclin-CDK complex, MPF (which stands for "mitotic promoting factor" or "maturation promoting factor"), allows the cell to proceed from the G2 checkpoint into mitosis. At the G2 checkpoint, MPF will phosphorylate a variety of proteins, triggering mitosis.

Animal cells contain at least four types of cyclins: G1-phase cyclin, G1/S-phase cyclin, S-phase cyclin, and M-phase cyclin. These types of cyclins will form CDK

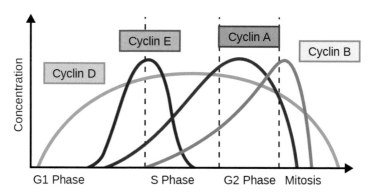

Fig. 2.9 Timeline of cyclin expression throughout the cell cycle. Different cyclins are expressed at different stages during the cell cycle. As the different CDKs require specific cyclins to function this is a way of regulating CDK activity during the cell cycle. For example, cyclin E is produced in the late G1 and early S phase

complexes in relation to their name, for example, S-phase cyclin will bind to its specific CDK and form S-CDK.

The CDKs are constantly expressed throughout the entire cycle. However, cyclin is produced and destroyed at various rates in different cell cycle phases. This rise and fall in cyclin levels helps control the cell cycle by modulating which active CDKs are present (Fig. 2.9).

Each cyclin-CDK complex phosphorylates a different set of proteins. S-CDK catalyzes phosphorylation of the protein that initiates DNA replication. M-CDK phosphorylates condensing proteins, which are essential for chromosome condensation, as well as lamin proteins, which form the nuclear membrane network. M-CDK also stimulates the assembly of the mitotic spindle by phosphorylating proteins required for microtubule formation.

Alongside the cyclin-CDK complex, other extracellular signals regulate cell size and cell number. They can be divided into three major classes:

- **Mitogens:** stimulate cell division by triggering G1/S-CDK.
- **Growth factor:** promotes cell growth by triggering protein synthesis and other macromolecules, and prevents their degradation.
- **Survival factor:** promotes cell survival by suppressing apoptosis (programmed cell death).

Microscopy

Scientists use microscopes (micro- = "small"; -scope = "to look at") to study structures that are too small to see by eye, like most cells. Using microscopes it is possible to observe the various stages of mitosis in action.

Light Microscopy

Light microscopy is the most commonly applied microscopy technique and uses light from the visible region of the spectrum. It often requires staining of the specimen to be able to visualize the structures of interest.

Light microscopy is limited to a minimum resolution of approximately 200 nm. The minimum resolution is defined as the distance between two points that are still distinguishable as two separate entities. The resolution is limited by the physical properties of light and the lens of the microscope.

Fluorescence Microscopy

Fluorescence microscopy is a subset of light microscopy. Rather than relying on traditional visible light sources, fluorescence microscopy uses higher wavelength light sources and takes advantage of the different emission and transmission wavelengths of fluorophores to produce high contrast images.

Briefly, structures of interest are labeled with a fluorescent dye; high wavelength light is shone onto the sample which excites the fluorescent dye, causing it to emit light of a different wavelength. The microscope is able to capture this emitted light, thus showing where structures are located within a cell. Through the use of filters and precise light sources such as lasers, it is possible to generate very sharp, high contrast images of at high magnification. Furthermore, by specifically labeling structures in the cell we can more easily observe their localization and movement, both of which are key to understanding the cell cycle.

Electron Microscopy

Electron microscopes are capable of imaging at extremely high magnifications beyond the optical limits of light microscopes, due to the high energy and short wavelength of free electrons. These microscopes are great at observing the molecular mechanisms of the cell cycle due to their extreme magnification.

- **Transmission electron microscopes (TEMs):** use a high voltage electron beam to produce an image, and the electrons that pass through the sample are detected.
- **Scanning electron microscopes (SEMs):** produce an image by probing the surface of a specimen.
- **Reflection electron microscopes (REMs):** use an electron beam that is pointed at a thin specimen to detect the scattered electrons that are reflected by the specimen.

Paclitaxel

In the Mitosis simulation, you will study the effect of the plant poison paclitaxel on the mammalian cell cycle. Paclitaxel was first isolated from the bark of Pacific yew (*Taxus brevifolia*). The precursor of the drug can be synthesized easily from the extract of the leaves of the European yew (*Taxus baccata*) (Fig. 2.10). Most parts of the tree are toxic, except the bright red aril surrounding the seed. Paclitaxel interferes with the normal breakdown of microtubules during cell division, so

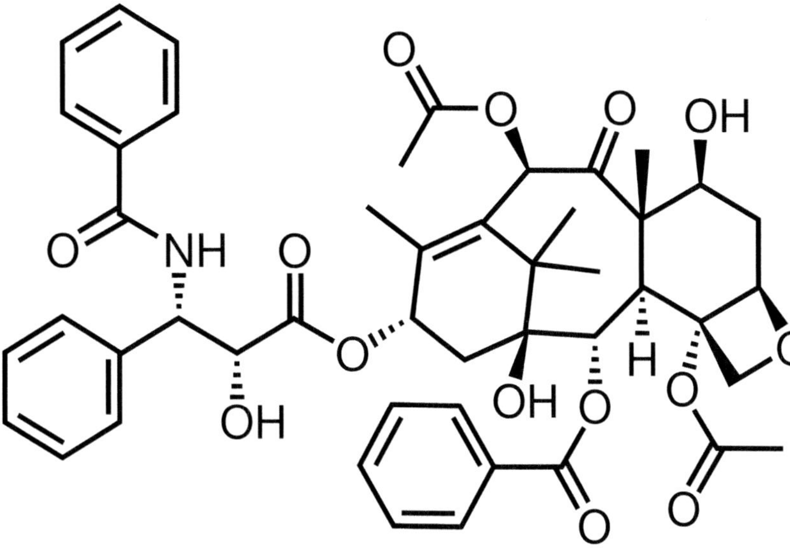

Fig. 2.10 Chemical structure of paclitaxel. Paclitaxel is a molecule which can be isolated from various species of yew. In humans, paclitaxel interferes with the breakdown of microtubules during cell division, thus disrupting the cell cycle. For this reason it has found use as a cancer therapeutic

disrupting the cell cycle. As such Paclitaxel is now used to treat a number of types of cancers where the cell cycle has become deregulated.

2.3 Let's Get Started

You are now equipped with all the knowledge you will need to successfully complete the Mitosis simulation. Let's go straight to the virtual lab and visualize mitosis using various microscopes and see if you can inhibit or accelerate it using an exciting new compound. Can you think how this compound could be used therapeutically?

Techniques Used in the Lab
- Microscopy

Learning Objectives

At the end of this simulation, you will be able to...

- Visualize the basic structures of eukaryotic cells including the main cellular components and DNA packaging via immersive animations
- Describe the key characteristics of the cell cycle's different stages: interphase (G1, S and G2) and mitosis
- Use different microscopy techniques to observe the five phases of the mitosis and identify their main characteristics
- Compare the cell cycle checkpoints and the molecules that control them (cyclins and cyclin-dependent kinases)

ACCESS THE VIRTUAL LAB SIMULATION HERE www.labster.com/springer **BY USING THE UNIQUE CODE AT THE END OF THE PRINTED BOOK. IF YOU USE THE E-BOOK YOU CAN PURCHASE ACCESS TO THE SIMULATIONS THROUGH THE SAME LINK.**

Further Reading

Alberts B et al (2015) The Molecular Biology of the Cell, 6th edn. Garland Science, Abingdon

Besson A, Dowdy SF, Inhibitors RJMCDK (2008) Cell Cycle Regulators and Beyond. Dev Cell 14(2):159–169

Jordan MA et al (1996) Mitotic Block Induced in HeLa Cells by Low Concentrations of Paclitaxel (Taxol) Results in Abnormal Mitotic Exit and Apoptotic Cell Death. Cancer Res 56(4):816–825

Kastan MB, Bartek J (2004) Cell-cycle checkpoints and cancer. Nature 432: p:316

OpenStax, Biology. OpenStax CNX. Jun 1, 2018. http://cnx.org/contents/185cbf87-c72e-48f5-b51e-f14f21b5eabd@11.2

Roukos V et al (2015) Cell cycle staging of individual cells by fluorescence microscopy. Nat Protoc 10: p:334

Urey LA et al (2014) Campbell Biology, 10th edn. Pearson, Boston

Wilson CR, Sauer JM, Taxines HSB (2001) a review of the mechanism and toxicity of yew (Taxus spp.) alkaloids. Toxicon 39(2):175–185

Meiosis

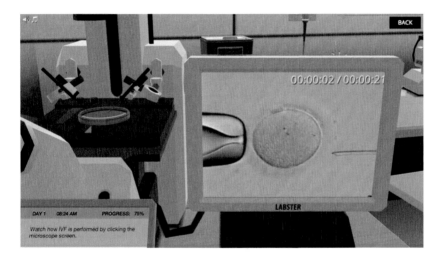

3.1   Meiosis Simulation

In the Meiosis simulation you will learn about the second major type of cell division which gives rise to the haploid male and female gametes. Learn about the various stages of meiosis and how they compare to those of mitosis before learning about *in vitro* fertilization and preimplantation genetic diagnosis. Will you be able to apply this knowledge to help a couple who are hoping to have a baby through *in vitro* fertilization?

Identify Male and Female Gametes
In the Meiosis lab, you will learn to identify the main characteristics of male and female gametes: the sperm and ovum. You will use a light microscope to identify the gametes and their key structures (Fig. 3.1). During this process, your knowledge of the basic principles of meiosis and fertilization will be tested.

Learn About the Stages of Meiosis
Meiosis comprises of a series of stages similar to mitosis, but with some significant differences. You will use the lily as a model organism to observe the stages of meiosis and their characteristics. Both mitosis and meiosis are types of cell division. In this lab, you will learn the differences between them (Fig. 3.2).

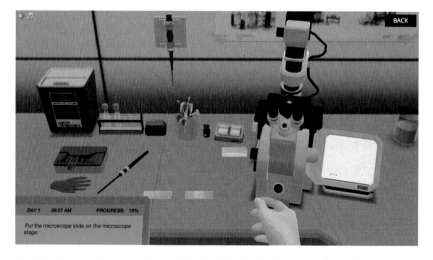

Fig. 3.1 Observe the stages of meiosis in the Meiosis simulation

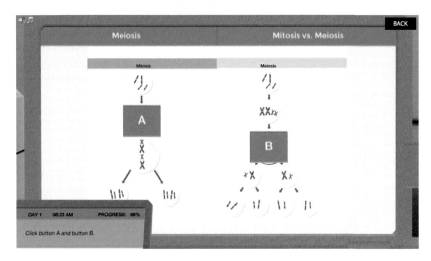

Fig. 3.2 Mini-game in the Meiosis simulation to learn about the differences between meiosis and mitosis

Fig. 3.3 Help a young couple give birth to a healthy baby in the Meiosis simulation

Learn About *In Vitro* Fertilization

After you have prepared the sperm and ovum, you will learn how *in vitro* fertilization is performed. Prior to embryo implantation, you will perform a diagnostic test to ensure the embryo has no chromosomal abnormalities. Will you be able to help your patients give birth to a healthy baby (Fig. 3.3)?

3.2 Meiosis Theory Content

In the Meiosis simulation you will explore the details of meiosis, beginning by examining human gametes under the microscope and finishing by helping a couple who are hoping to have a baby via *in vitro* fertilization. The Meiosis theory content covers all the information you will need to perform these and other experiments when in the lab.

Cell Reproduction

Mitosis and meiosis are both involved in reproduction; however, they both play different roles. Asexual reproduction only involves mitosis, while sexual reproduction involves both mitosis and meiosis.

A somatic cell is any cell of the body with the exception of the gametes. Somatic cells contain two sets of chromosomes, which are found in homologous pairs, with one chromosome from each pair coming from one parent. Humans have a total of 46 chromosomes, 23 coming from the mother and 23 from the father. Of these 23, 22 (the autosomes) are described as homologous pairs as they are of similar size and appearance. The exception is the 23rd pair, the sex chromosomes (X and Y) where the Y chromosome is much smaller than the X.

Each chromosome consists of two sister chromatids, attached to each other by a protein complex called cohesin which binds the two DNA strands forming a region known as the centromere.

Non-sister chromatids are any two chromatids in a pair of homologous chromosomes. In other words, two maternal chromatids would be described as sisters, whereas a paternal and maternal chromatid would be described as non-sister chromatids.

Unlike diploid somatic cells, gametes contain only a single set of chromosomes. They are therefore called haploid. They are formed by meiosis, a specialized type of cell division, which reduces the number of chromosomes by half.

The only haploid cells in humans are the gametes or the sperm and ovum (egg). During reproduction, they fuse to form a diploid zygote, the precursor of the embryo. In this way, DNA from both the mother and father can be mixed to create a unique new individual.

Meiosis

Meiosis, unlike mitosis, creates daughter cells that have half as many chromosomes as the parent cell (Fig. 3.4).

Meiosis is preceded by an interphase consisting of the G1, S, and G2 phases, which are nearly identical to the same phases in mitosis. During DNA duplication in the S phase, each chromosome is replicated, producing two identical copies called sister chromatids, that are held together at the centromere by cohesin proteins. Cohesin holds the chromatids together until anaphase II.

The function of meiosis is to reduce the number of chromosomes from diploid to haploid, thus allowing for paternal and maternal DNA to mix upon fertilization, creating genetic diversity. Meiosis consists of two nuclear divisions, meiosis I and meiosis II, that reduce the number of chromosomes in preparation for sexual reproduction. The stages of meiosis I and II are similar to those seen in mitosis, but the outcome is very different.

Meiosis I

Meiosis I consists of the following stages:

Prophase I
Early in prophase I, before the chromosomes can be seen clearly microscopically, the homologous chromosomes are attached by their tips to the nuclear envelope by proteins. As the nuclear envelope begins to break down, the proteins associated with homologous chromosomes bring the pairs close to each other. Importantly you should remember that, in mitosis, homologous chromosomes do not pair together.

The tight pairing of the homologous chromosomes is called synapsis. During synapsis, the genes on the chromatids of the homologous chromosomes are aligned precisely with each other. The synaptonemal complex supports the exchange of chromosomal segments between non-sister homologous chromatids, a process

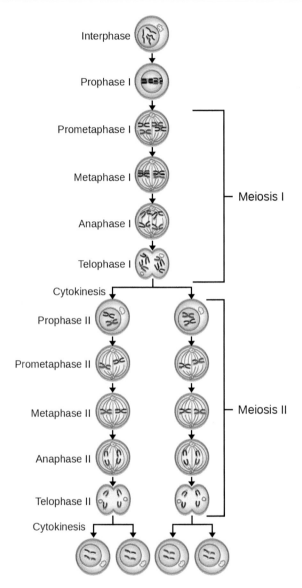

Fig. 3.4 Progression of meiosis. Meiosis proceeds in two steps. During meiosis I two haploid cells are formed each containing one set of homologous chromosomes. During meiosis II the sister chromatids of these homologous pairs are separated, producing four gametes

called crossing over. Crossing over can be observed visually after the exchange as chiasmata (singular = chiasma).

Crossover occurs between non-sister chromatids of homologous chromosomes. The result is an exchange of genetic material between homologous chromosomes. In effect shuffling the alleles on parental chromosomes, so that the gametes carry combinations of genes different from either parent.

Prometaphase I
The key event in prometaphase I is the attachment of the spindle fiber microtubules to the kinetochore proteins at the centromeres.

Metaphase I
During metaphase I, the homologous chromosomes are arranged in the center of the cell with the kinetochores facing opposite poles.

Anaphase I
In anaphase I, the microtubules pull the linked chromosomes apart. The sister chromatids remain tightly bound together at the centromere.

Telophase I and Cytokinesis
In telophase, the separated chromosomes arrive at opposite poles, these are separated into individual cells during cytokinesis.

Two haploid cells are the end result of the first meiotic division. The cells are haploid because at each pole there is just one of each pair of the homologous chromosomes. Therefore, only one full set of chromosomes is present. This is why the cells are considered haploid; there is only one chromosome set, even though each homolog still consists of two sister chromatids. Recall that sister chromatids are merely duplicates of one of the two homologous chromosomes (except for changes that occurred during crossing over). In meiosis II, these two sister chromatids will separate, creating four haploid daughter cells.

Meiosis II

During meiosis II, the sister chromatids within the two daughter cells separate, forming four new haploid gametes.

Prophase II
If the chromosomes de-condensed in telophase I they condense again. If nuclear envelopes were formed, they fragment into vesicles. The centrosomes that were duplicated during interkinesis move away from each other toward opposite poles, and new spindles are formed.

Prometaphase II
The nuclear envelopes are completely broken down, and the spindle is fully formed. Each sister chromatid forms an individual kinetochore that attaches to microtubules from opposite poles.

Metaphase II
The sister chromatids are maximally condensed and aligned at the equator of the cell.

Anaphase II
The sister chromatids are pulled apart by the kinetochore microtubules and move toward opposite poles. Non-kinetochore microtubules elongate the cell.

Telophase II and Cytokinesis
The chromosomes arrive at opposite poles and begin to decondense. Nuclear envelopes form around the chromosomes. Cytokinesis separates the two cells into four unique haploid cells. At this point, the newly formed nuclei are both haploid. The cells produced are genetically unique because of the random assortment of paternal and maternal homologs and because of the recombining of maternal and paternal segments of chromosomes (with their sets of genes) that occurs during crossing over.

Mitosis and Meiosis Comparison

The two types of cell division, mitosis and meiosis, serve different purposes. Mitosis is used for cell divisions where chromosomes/DNA are exactly replicated in

Fig. 3.5 Comparison of mitosis and meiosis. While some aspects of mitosis and meiosis appear similar the eventual outcomes are completely different. Mitosis results in two identical diploid daughter cells. Whereas, the two rounds of meiosis result in four genetically diverse haploid cells
▶

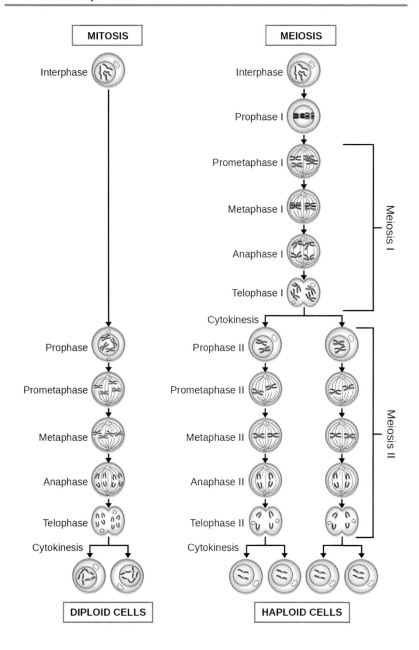

new cells. Whereas, meiosis reduces the number of chromosomes in new cells, an essential step if genetic information from each parent is to be contributed to their offspring (Fig. 3.5).

Nondisjunction

The above descriptions cover the process of meiosis when it happens correctly. However, occasionally entire chromosomes, or fragments of chromosomes, can be duplicated or lost resulting in an altered chromosome number. This occurs during meiosis when pairs of homologous chromosomes fail to separate correctly during meiosis and is termed nondisjunction (Fig. 3.6).

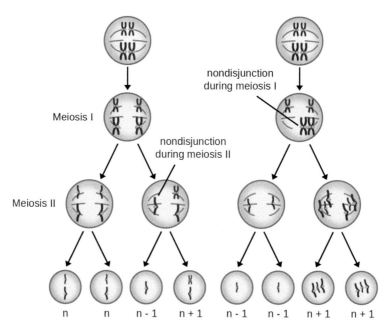

Fig. 3.6 Outcomes of nondisjunction. If nondisjunction occurs during meiosis I both chromosomes of one homologous pair migrate to the same pole. This means that following meiosis II, two of the gametes will have an extra chromosome and two will lack a chromosome. If nondisjunction occurs during one arm of meiosis II then two gametes will be normal, one will have an extra chromosome and one will lack a chromosome

Nondisjunction can occur during either meiosis I or II, with different results:

Nondisjunction Meiosis I
If nondisjunction occurs in meiosis I, both chromosomes of a homologous pair migrate to the same pole, leaving one daughter cell without any of this chromosome. The cell then goes through meiosis II normally. The products are four gametes: two of which have an extra chromosome (n + 1), with the other two gametes lacking a chromosome (n − 1).

Nondisjunction Meiosis II
If nondisjunction occurs in meiosis II, both sister chromatids of a chromosome migrate to the same pole of the cell correctly. However, during meiosis II if one chromosome separates abnormally the products will be four gametes: two being normal, one with an extra chromosome (n + 1) and one lacking a chromosome (n − 1).

Genetic Consequence of Meiosis

Meiosis produces cells that are genetically unique to the recombination of maternal and paternal segments during that occurs during crossing over, and because of the random assortment of these chromosomes into the gametes (Fig. 3.7).

Chromosome Number Disorders

Nondisjunction can lead to alterations in chromosome number, which in turn can often lead to severe developmental disorders. Individuals with the correct chromosome number are referred to as euploid, so in humans this would be 22 pairs of autosomes and one pair of sex chromosomes. Those with an incorrect chromosome number are referred to as aneuploid, a term that includes monosomy (the complete loss of one chromosome) or trisomy (the complete gain of an extra chromosome) and other variations.

Monosomic human zygotes missing any one copy of an autosome invariably fail to develop to birth because they have only one copy of essential genes. However, sex chromosomes can be more forgiving; for example, Turner syndrome is a chromosome disorder which affects females. Those with Turner syndrome carry only one copy of the X chromosome, and while the disorder is associated with some developmental issues individuals are commonly able to live healthy lives.

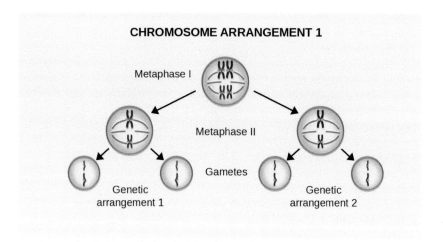

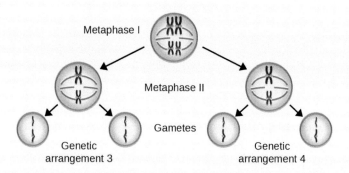

Fig. 3.7 Genetic outcomes of meiosis. Let's use an example image with two chromosomes ($n = 2$) undergoing independent assortment at metaphase I. In this case there are two possible arrangements, the paternal and maternal chromosomes could separate independently as seen in the upper panel, or the maternal and paternal chromosomes could mix as shown in the lower panel. This ultimately means that there are four possible arrangements. Now imagine how many possibilities there are with 23 chromosomes ($n = 23$)

Most autosomal trisomies also fail to develop to birth; however, duplications of some of the smaller chromosomes (13, 15, 18, 21, or 22) can result in children who can survive for several years, albeit with often severe developmental disorders. Down syndrome is perhaps the most well known viable trisomy whereby individuals carry an extra copy of chromosome 21.

In Vitro Fertilization

In the Meiosis simulation, you will perform an *in vitro* fertilization (IVF) experiment, the most commonly used assisted reproductive technology (ART). Other types of ART involve surgically removing eggs from a woman's ovaries, combining them with sperm in the laboratory, and returning them to the woman's body or donating them to another woman.

To test the result of ART, a preimplantation genetic diagnosis is often performed.

Preimplantation Genetic Diagnosis

Preimplantation genetic diagnosis (PGD) is a reproductive technology used within an IVF cycle which is used to diagnose for the presence of several genetic disorders in early embryos prior to implantation and pregnancy.

The process of IVF and PGD is as follows:

- Provide hormone treatment to stimulate the patient's ovaries
- Collect eggs from the patient
- Fertilize eggs in the laboratory with paternal sperm
- Eggs that are successfully fertilized divide and multiply to form a developing embryo called a zygote
- After three days, the developing embryo will contain about eight cells known as blastomeres
- Remove one cell for genetic testing, the remaining cells will continue to grow and the embryo should be unharmed

The most commonly applied method for PGD is fluorescence *in situ* hybridization (FISH). The embryo cells are fixed onto a microscope slide and hybridized with DNA probes. Each of these probes is specific for a particular part of a chromosome and are labeled with a distinct fluorochrome. The cell can then be visualized under a microscope, if a particular fluorescent signal is missing, or present more frequently, then it suggests that there is a potential issue with the genetic health of the embryo.

3.3 Let's Get Started

Meiosis is a fundamental process driving the wonderful variation you can see in the individuals around you. Every one of us is genetically unique and this is because of meiosis and the subsequent combination of your parents' gametes. You should now know everything you need to complete the Meiosis simulation successfully. Will you be able to use this knowledge to help a couple give birth to a healthy baby?

Techniques Used in the Lab
- Microscopy
- *In vitro* fertilization

Learning Objectives
At the end of this simulation, you will be able to…

- Explain assisted reproduction technology and preimplantation genetic diagnosis
- Define the basic principle of meiosis
- Understand the main differences between mitosis and meiosis
- Describe how nondisjunction can lead to genetic disorders
- Use the microscope to observe the phases of meiosis and identify their main characteristics

ACCESS THE VIRTUAL LAB SIMULATION HERE www.labster.com/ springer BY USING THE UNIQUE CODE AT THE END OF THE PRINTED BOOK. IF YOU USE THE E-BOOK YOU CAN PURCHASE ACCESS TO THE SIMULATIONS THROUGH THE SAME LINK.

Further Reading

Alberts B et al (2015) The Molecular Biology of the Cell, 6th edn. Garland Science, Abingdon

Openstax, Biology. OpenStax CNX. Jun 1, 2018. http://cnx.org/contents/185cbf87-c72e-48f5-b51e-f14f21b5eabd@11.2

Robinson WP, McFadden DE (2002) Chromosomal genetic disease: numerical aberrations. Encyclopedia of Life Sciences. John Wiley & Sons Ltd, pp 1–7. www.els.net

Sermon K, Van Steirteghem A, Liebaers I (2004) Preimplantation genetic diagnosis. Lancet 363(9421):1633–1641

Urey LA et al (2014) Campbell Biology, 10th edn. Pearson, Boston

Cellular Respiration 4

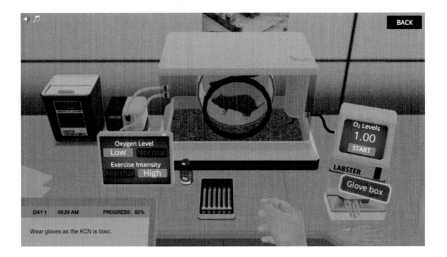

© Labster ApS under license to Springer Verlag GmbH 2018
S. Stauffer et al., *Labster Virtual Lab Experiments: Basic Biology*,
https://doi.org/10.1007/978-3-662-57996-1_4

4.1   Cellular Respiration Simulation

All living organisms require energy to survive and thrive. In the Cellular Respiration simulation you will learn all about the smart ways your cells get the energy you need to keep moving, breathing and thinking. You will then use that knowledge to help your favorite soccer team with a new and improved training plan. Finally, you will perform respiratory experiments to understand how your body utilizes the energy and how that process can be optimized.

Design Your Experiment
As a scientist, one of the most important tasks you have is designing your experiments correctly. In this simulation, you will learn how to make decisions about the best experimental strategies, which animal models you want to use and which experimental observations to focus on (Fig. 4.1). Measuring energy consumption is not an easy task, and cellular respiration is a complex metabolic route. But no worries, your colleague Marie will join you in the simulation and help you out!

Perform Respirometry Experiments
You will perform respirometry experiments with mice, where you can control variables like how fast they run and the oxygen levels they are exposed to. You will

Fig. 4.1   Respirometry experimental setup in the Cellular Respiration simulation

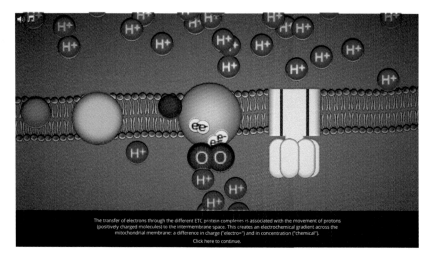

The transfer of electrons through the different ETC protein complexes is associated with the movement of protons (positively charged molecules) to the intermembrane space. This creates an electrochemical gradient across the mitochondrial membrane: a difference in charge ("electro=") and in concentration ("chemical").
Click here to continue.

Fig. 4.2 Electron transport chain animation in the Cellular Respiration simulation

use a respirometer to measure oxygen consumption, as well as a chip that will give you information about lactate and glucose blood levels in real time. You will also use cyanide, as it can help you understand how adenosine triphosphate (ATP) production depends on the oxidative phosphorylation process. And don't worry, the mice are virtual and won't be harmed.

Your colleague Marie will guide you through the most important metabolic routes using detailed animations (Fig. 4.2) to make it easier to visualize the complex molecular components such as the electron transport chain proteins or electrochemical gradients.

Analyze Your Data and Draw Conclusions

After all the hard work, it's time to take a look at the data you've gathered and draw conclusions (Fig. 4.3). As the final part of the Cellular Respiration simulation, you will use your results to write a report with advice for the coach of your favorite soccer team.

Does it have an impact on the team's achievements? And will you help them win the world cup this year?

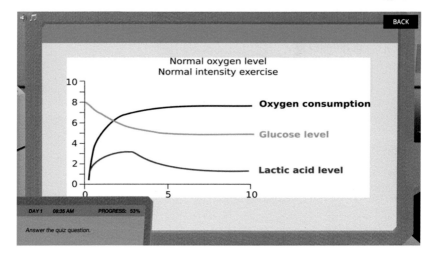

Fig. 4.3 Measurement of key respiration markers in the Cellular Respiration simulation

4.2   Cellular Respiration Theory Content

Energy is a fundamental requirement for life. As mammals, we generate energy from the food we eat and the air we breathe through a mechanism known as cellular respiration. The cellular respiration theory content below will take you through the basics of this key process allowing you to help your favorite soccer team in the cellular respiration simulation.

Metabolism

The word metabolism comes from the Greek word "metabolë" meaning "change". It refers to the chemical transformations that take place in cells. These reactions sustain all life, i.e. allow the organism to maintain their structures, grow, reproduce and respond to their environment.

Metabolic reactions can be divided into catabolic reactions or anabolic reactions, both types of reactions are interconnected and form metabolic pathways (Fig. 4.4).

Catabolic reactions: break down larger complex molecules into smaller molecules. These reactions create the building blocks for creating other molecules

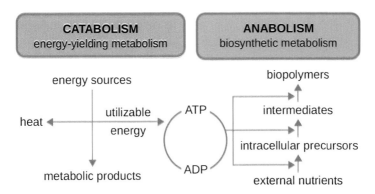

Fig. 4.4 **Catabolism, anabolism, and ATP.** Catabolism and anabolism are two parts of the metabolic process. Catabolism refers to the break down of larger products into their small constituents and is also responsible for generating utilizable energy in the form of ATP. Anabolism refers to the construction of larger molecules such as proteins or DNA

and are the types of reactions that lead to the synthesis of energy in the form of ATP. Multiple catabolic pathways converge in cellular respiration.

Anabolic reactions: utilize the energy stored in ATP to build large molecules, such as proteins or DNA.

The main connection between metabolic pathways is that catabolic reactions produce energy or ATP which is then used by anabolic reactions.

Energy Use

Our bodies require energy, but also need to be able to store and direct that energy. ATP is the ideal molecule for this and can be thought of as the energy currency, to be spent where it is required in the body. Our cells can produce ATP via ATP synthesis, which requires the oxidation of glucose via aerobic cellular respiration or lactic acid fermentation, depending on oxygen availability. When we exercise, our energy requirements increase and glucose is consumed more quickly. Also, oxygen consumption of the electron transport chain increases resulting in heavy breathing. We can measure oxygen consumption using respirometry.

In situations with low oxygen for prolonged periods of time, lactic acid can accumulate leading to lactic acidosis.

The importance of aerobic respiration is reflected in the consequences of blocking it: cells are not able to produce the energy our bodies need to maintain vital functions. This can even lead to the death of the organism. On the other hand, boosting aerobic respiration, for example by increasing oxygen availability through blood doping, leads to an increase in energy production used illegaly to improve the performance of athletes.

ATP

ATP is a molecule that acts as a universal energy currency for living cells. Its structure consists of the nucleoside adenosine and a tail of three phosphate groups (Fig. 4.5).

During ATP synthesis energy is safely stored as chemical energy in the structure of ATP, specifically in the high energy phosphate bonds. The negative charges in the phosphate groups repel each other and need high amounts of energy to bond them together. When these high-energy bonds are broken, this energy is released through ATP hydrolysis.

Cellular Respiration

Cellular respiration is the process by which animals convert food into a type of energy usable by their cells, known as ATP. The first step of cellular respiration is called glycolysis and results in the formation of pyruvate.

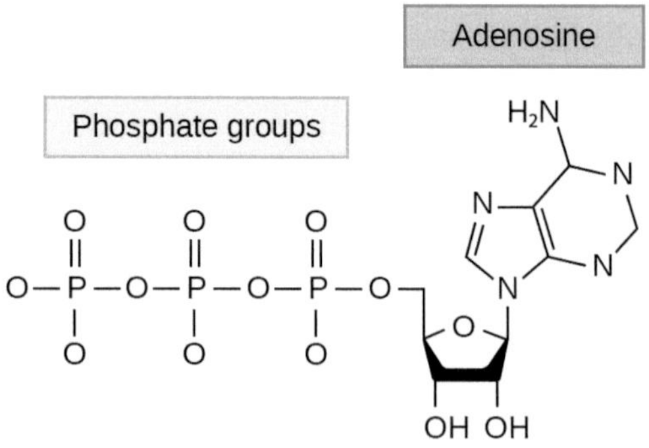

Fig. 4.5 Chemical structure of ATP. ATP stores energy in its phosphate bonds, these bonds require a lot of energy to form as phosphate groups naturally repel each other. When broken they release energy which can be used by a variety of cellular processes

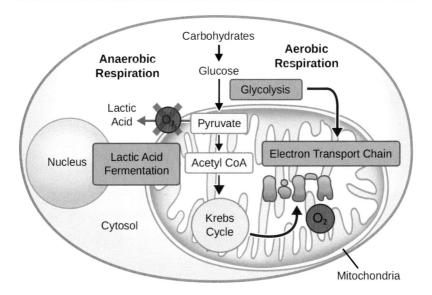

Fig. 4.6 Cellular respiration processes. When oxygen is present, aerobic cellular respiration can occur. The pyruvate molecules generated by glycolysis enter the Krebs cycle and electron transport chain generating a large amount of ATP. In the absence of ATP pyruvate is converted into lactic acid via anaerobic respiration, which results in the formation of a lower amount of ATP

Aerobic cellular respiration occurs when oxygen is present, and pyruvate will enter the Krebs cycle allowing the electron transport chain to proceed. Anaerobic cellular respiration does not require the presence of oxygen, and pyruvate will undergo lactic acid fermentation. Comparing the result of aerobic and anaerobic respiration highlights why oxygen is so important for cellular respiration (Fig. 4.6).

Aerobic Cellular Respiration

When oxygen is present, aerobic cellular respiration takes place. This process can be divided into three main phases: glycolysis, the Krebs cycle (sometimes called the tricarboxylic acid, TCA cycle, or citric acid cycle), and the electron transport chain. As by-products, CO_2 and H_2O are produced, which leave the body via respiration, urination, and perspiration (Fig. 4.7).

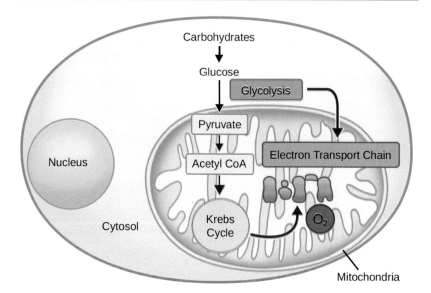

Fig. 4.7 Aerobic respiration. In the presence of oxygen pyruvate generated by glycolysis enters the Krebs cycle and electron transport chain generating a large amount of ATP molecules. One glucose molecule is ultimately converted into six CO_2 molecules and produces up to 38 ATP molecules, as detailed in Table 4.1

The general equation for the cellular respiration process is (detailed in Table 4.1):

$$Glucose\ (C_6H_{12}O_6) + 6\ O_2 + 38\ ADP \rightarrow 6\ C_2O + 6\ H_2O + 38\ ATP + heat$$

Glycolysis

Glycolysis is the first step in cellular respiration and occurs in the cytoplasm of the cell. The word glycolysis literally means "breaking down sugars". During this process, the 6-carbon glucose molecule is split into two 3-carbon pyruvate molecules, producing only a small amount of energy. Nearly all living organisms carry out glycolysis as part of their metabolism. The process does not use oxygen and is therefore anaerobic.

Table 4.1 **Theoretical maximum yields of ATP during cellular respiration.** Table detailing reactions and molecules produced during aerobic respiration. In total, a single glucose molecule is converted into six CO_2 molecules yielding a theoretical maximum of 38 ATP molecules

Source	Carbon Flow	Reduced Coenzymes	Substrate-level Phosphorylation $\downarrow$ Net ATP	Oxidative Phosphorylation $\downarrow$ Net ATP	Theoretical Maximum Yield of ATP
Glycolysis (EMP)	Glucose (6C) $\downarrow$ 2 pyruvates (3C)	2 NADH	2 ATP	6 ATP from 2 NADH	8
Krebs Cycle Preparation	2 pyruvates (3C) $\downarrow$ 2 acetyl (2C) + 2 CO_2	2 NADH		6 ATP from 2 NADH	6
Krebs Cycle	2 acetyl (2C) $\downarrow$ 4 CO_2	6 NADH 2 FADH$_2$	2 ATP	18 ATP from 6 NADH 4 ATP from 2 2 FADH$_2$	24
Total:	Glucose (6C) $\downarrow$ 6 CO_2	12 NADH 2 FADH$_2$	4 ATP	34 ATP	38

Krebs Cycle

The Krebs cycle is the continuation of aerobic cellular respiration after glycolysis, and it takes place in the mitochondrial matrix. During the Krebs cycle preparation step, pyruvate is transformed into acetyl-CoA. After this, several Krebs cycle reactions couple the oxidation of pyruvate to CO_2, reduce the electron carriers NAD^+ and FAD, and produce ATP via substrate-level phosphorylation. The reduced electron carriers (NADH and FADH$_2$) are used in the electron transport chain to produce more ATP by oxidative phosphorylation.

Electron Transport Chain

The electron transport chain takes place in the inner mitochondrial membrane and is the final step in aerobic cellular respiration (Fig. 4.8). It consists of a series of

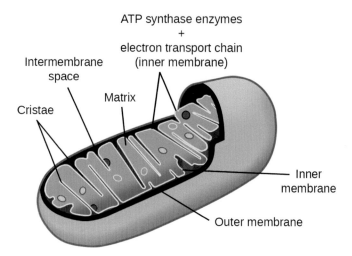

ATP synthase enzymes
+
electron transport chain
Intermembrane   (inner membrane)
space
                        Matrix
Cristae

                                                          Inner
                                                          membrane

                                            Outer membrane

Fig. 4.8 Structure of a mitochondrion. The mitochondrion is the powerhouse of the cell using the electron transport chain and oxidative reactions to generate large amounts of ATP

redox reactions that transfer electrons from NADH and $FADH_2$ through various intermediates to the final electron acceptor, oxygen. This process generates an electrochemical gradient that couples the oxidative reactions with the phosphorylation of ADP producing ATP in a process called oxidative phosphorylation.

For a detailed overview of the electron transfer chain see Fig. 4.9. The electron transport chain consists of four protein complexes (labeled I to IV), two mobile electron carriers, ubiquinone and cytochrome C (Q and Cyt C) and ATP synthase.

Chemiosmosis

Chemiosmosis is the movement of ions across a selectively permeable membrane down an electrochemical gradient. In the context of cellular respiration, it refers to the diffusion of protons through the ATP synthase. This process is coupled to the generation of ATP via oxidative phosphorylation. If the electrochemical gradient is disrupted for any reason, the electron transport chain is stopped.

Different electron carriers yield different amounts of ATP. On average, each NADH molecule results in the production of three ATP molecules, and each $FADH_2$ molecule produces two ATP molecules.

Intermembrane space

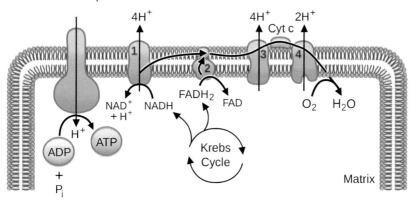

Fig. 4.9   The electron transport chain in detail. Complex I (*1* in the Fig.) and II (*2* in the Fig.) accept electrons from the electron carriers, NADH and FADH$_2$, produced during glycolysis and the Krebs cycle. The electrons flow through ubiquinone, cytochrome C and Complex III (*3* in the Fig.) before reaching Complex IV (*4* in the Fig.) where they reduce molecular oxygen producing in a water molecule. The energy released through the oxidation of NADH and FADH2 by Complex I and II enables the formation of an electrochemical proton gradient across the inner mitochondrial membrane. This electrochemical gradient is utilized by ATP synthase to produce ATP via chemiosmosis

Anaerobic Cellular Respiration

Anaerobic cellular respiration is the only source of ATP when insufficient oxygen is available for aerobic respiration.

Anaerobic respiration takes place after glycolysis. This process differs among species, and in mammals it is called lactic acid fermentation. Lactic acid fermentation does not require oxygen to proceed, although it can occur in its presence. During lactic acid fermentation, pyruvate is converted into lactic acid and the NADH produced during glycolysis is recycled. The net production of energy comes from glycolysis in the form of two ATP molecules, much fewer than in aerobic respiration; however, the reaction is much quicker (Fig. 4.10).

Fig. 4.10 Anaerobic respiration chemical reaction. In the absence of oxygen, pyruvate generated by glycolysis is converted into lactic acid. The overall anaerobic reaction generates two ATP molecules but does so more quickly than aerobic respiration

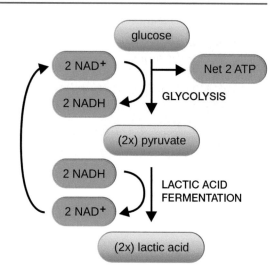

Respirometry

In the Cellular Respiration simulation, you will perform respirometry experiments to follow the processes described in this chapter. The term respirometry refers to a group of techniques where a quantitative measurement of respiration is performed. It is based on an indirect measurement of metabolic changes by recording variations in oxygen levels in a respirometer due to energy consumption in an experimentally controlled model.

4.3   Let's Get Started

Nearly every cell in your body is its own little power plant, with a little sugar and some oxygen the mitochondria in your cells can produce all the energy you need to move, breath and even think. Will you be able to use your new knowledge to optimize your bodies energy usage in the Cellular Respiration simulation?

Techniques Used in the Lab
- Respirometry

Learning Objectives
At the end of this simulation, you will be able to...

- Use respirometry experiments to achieve an integrated view of the cellular respiration process
- Explain the importance and good use of animal models in experimentation
- Describe the process of glycolysis and lactic acid fermentation and how they are interconnected with the Krebs cycle
- Draw a simple overview of the Krebs or Citric Acid cycle
- Describe an overview of how the electron transport chain works and the role of the redox electron carriers
- Explain the importance of oxidative phosphorylation, chemiosmosis, and electrochemical gradients for the electron transport chain
- Calculate the amount of ATP produced in the different metabolic routes

ACCESS THE VIRTUAL LAB SIMULATION HERE www.labster.com/springer BY USING THE UNIQUE CODE AT THE END OF THE PRINTED BOOK. IF YOU USE THE E-BOOK YOU CAN PURCHASE ACCESS TO THE SIMULATIONS THROUGH THE SAME LINK.

Further Reading

Alberts B et al (2015) The Molecular Biology of the Cell, 6th edn. Garland Science, Abingdon
Nelson DL, Cox MM, Lehninger AL (2013) Lehninger Principles of Biochemistry, 6th edn. WH Freeman, New York
OpenStax, Biology. OpenStax CNX. Jun 1, 2018. http://cnx.org/contents/185cbf87-c72e-48f5-b51e-f14f21b5eabd@11.2
Urey LA et al (2014) Campbell Biology, 10th edn. Pearson, Boston

Protein Synthesis

S. Stauffer et al., *Labster Virtual Lab Experiments: Basic Biology*,
https://doi.org/10.1007/978-3-662-57996-1_5

5.1 Protein Synthesis Simulation

In the Protein Synthesis simulation, you will learn about the difference between protein synthesis in prokaryotes and eukaryotes. You will see how we can use these systems to make artificial proteins to improve health, but also how these artificial proteins can be used for nefarious purposes such as blood doping. Will you be able to help the authorities identify the doping cheat?

Prepare Recombinant Erythropoietin and Use the Mass Spectrometer
Your first task in the lab will be to prepare recombinant erythropoietin (EPO) using bacteria and a mammalian cell culture system. You will then measure the mass to charge ratio using a mass spectrometer (Fig. 5.1).

Study the Translation Process from mRNA to Amino Acids
Making EPO in the lab requires a good understanding of translation. You will learn about the process from mRNA to amino acids and how amino acids are assembled to proteins; before seeing how we can use this system in the lab. A 3D animation will visualize how triplets of codons are translated into amino acids, how these amino acids are joined together by peptide bonds creating the primary

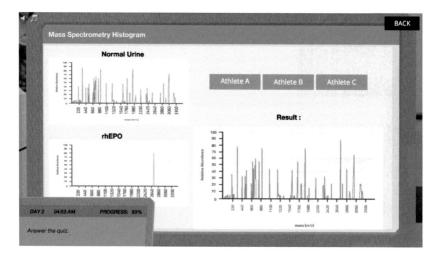

Fig. 5.1 Analyzing mass spectra in the Protein Synthesis simulation

Fig. 5.2 Animated EPO protein structure in the Protein Synthesis simulation

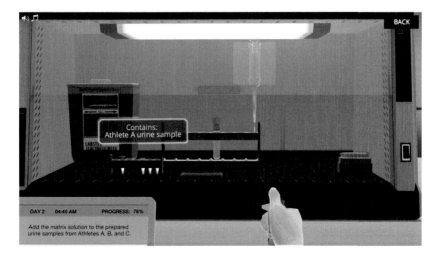

Fig. 5.3 Testing samples for doping in the protein synthesis simulation

protein structure, and furthermore, how the primary structure is folded into the secondary, tertiary and quaternary structures (Fig. 5.2).

Investigate Doping Among Professional Cyclists
In the last part of the Protein Synthesis simulation, you will use mass spectrometry and investigate if there are any athletes who are using rhEPO as a doping substance. You will do so by collaborating with the doping agent who collects urine samples during a large bicycle race (Fig. 5.3).
    Will you be able to detect if any of the athletes are doping?

5.2   Protein Synthesis Theory Content

The protein synthesis theory content will equip you with all the knowledge you need to complete the protein synthesis simulation. You will learn how to synthesize a recombinant protein called recombinant human erythropoietin (rhEPO), which is often used to treat anemic patients. However, rhEPO is also used for blood doping and you will employ mass spectrometry techniques to find which athletes have illicitly used rhEPO.

Expression System

The central dogma of molecular biology is an explanation of the flow of genetic information within a biological system as shown in Fig. 5.4. It refers to the flow of information from deoxyribonucleic acid (DNA) to ribonucleic acid (RNA) and finally into proteins. DNA to RNA conversion is carried out in a process called transcription. Then, the subsequent RNA sequence will be translated into protein in a process called translation.

Nucleic Acids: DNA and RNA

DNA and RNA represent the first step of an expression system, carrying the code that is used to design proteins. While both composed of nucleotide bases there are some significant structural differences as described below and shown in Fig. 5.5.

Fig. 5.4   Central Dogma of molecular biology. DNA to RNA transcription is mediated by the RNA polymerase and then, RNA to protein translation is mediated by the ribosome

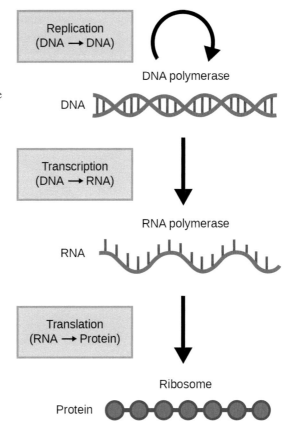

DNA

DNA is the hereditary material of all living organisms, including humans.

   DNA is a double helix made of two complementary strands. The information stored in DNA contains instructions for building and maintaining all the cells of an organism. DNA stores information as nucleotides; either A (adenine), T (thymine), C (cytosine) or G (guanine). Adenine (A) and thymine (T) or cytosine (C) and guanine (G) always pair up between the complementary strands. Nucleotides bind to each-other through the phosphate group attached to the 5th carbon of the sugar molecule, and the hydroxyl group attached to the 3rd carbon molecule forming a phosphodiester link. This allows us to apply a direction when we are describing DNA strands, designated with 5′ and 3′, respectively.

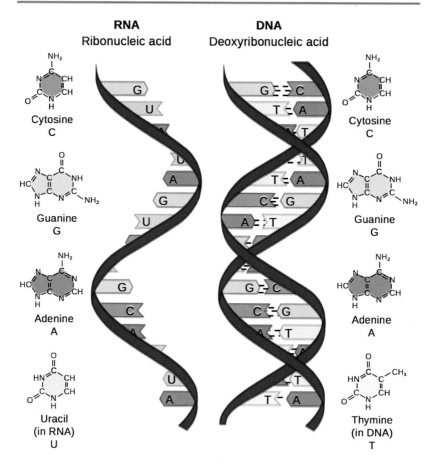

Fig. 5.5 Comparison of DNA and RNA. DNA and RNA are both polymeric molecules constructed from nucleotide bases. DNA can be thought of as the genetic storage mechanism, it is double-stranded with nucleotide bases paired between the two strands. RNA can be thought of as a more active genetic material, forming variants like mRNA or rRNA. Unlike DNA it is only single-stranded and uses uracil (U) in the place of thymine (T)

The weak hydrogen bonds between complementary nucleotides give DNA its characteristic double helix structure. The double helix is said to be antiparallel because one strand runs in the $5'{\rightarrow}3'$ direction and the other runs in the $3'{\rightarrow}5'$ direction.

The human genome consists of approximately 3.2 billion of these base pairs. Because of the enormous size of DNA, eukaryotes have developed an efficient DNA packaging system. DNA is found in the nucleus, chloroplasts, and mitochondria of eukaryotes. In prokaryotes, the DNA is not enclosed in a membranous envelope.

RNA
DNA is transcribed into RNA that stores information in a similar fashion to DNA; however, unlike DNA, RNAs are typically single-stranded molecules. Nucleotides on the single strand bind to each other using phosphodiester bonds as described above. As a single-stranded molecule, no hydrogen bonds are formed between strands and for this reason RNA is more susceptible to degradation or cleavage via enzyme activity.

RNA is mostly involved in protein synthesis. DNA molecules never leave the nucleus but instead use an intermediary to communicate with the rest of the cell. This intermediary is the messenger RNA (mRNA). Other types of RNA like ribosomal (rRNA), transfer (tRNA), and microRNA (miRNA) are mostly involved in protein synthesis and its regulation.

RNA, just like DNA, stores information in four bases: A (adenine), U (uracil), C (cytosine), and G (guanine). Unlike DNA, the thymine (T) bases are replaced by uracil (U).

Types of RNA

DNA can be thought of as the long-term storage solution for genetic material. Whereas RNA is more active in cellular functions and for that reason cells contain different types of RNA with many important functions in living organisms. The following are the most important types of RNAs:

mRNA (messenger RNA): is the product of protein-coding genes. It is produced in the nucleus and transported to the cytoplasm (in eukaryotes).

tRNA (transfer RNA): is a group of RNA molecules participating in the protein synthesis process by transporting amino acids. Several kinds of tRNA exist, each characterized by a specific anticodon in one end. The anticodon base will pair with a complementary codon in the mRNA, after which the corresponding amino acid carried by the tRNA is incorporated into the growing protein.

rRNA (ribosomal RNA): is the main component of the ribosomes and forms two subunits: a large subunit containing the 5S, 5.8S, and 28S molecules, and a small subunit containing the 18S molecule. The 28S and 18S molecules are easily recognized as distinct bands on a gel picture after electrophoresis with total RNA.

Regulatory RNAs: Apart from protein synthesis, RNAs also regulate gene expression and act as a cellular immune system. Examples of these are mRNAs and small interfering RNAs (siRNAs).

RNA is copied from its complementary DNA strand by an enzyme called RNA polymerase. RNA synthesis proceeds in the 5′ to 3′ direction. 5′ and 3′ refer to the number of carbon molecules in the sugar backbone. The 5′ carbon has a phosphate group, and the 3′ carbon has a hydroxyl group. The 5′ carbon binds to the phosphate group of the previous nucleotide leaving a negatively charged oxygen molecule free (Fig. 5.6).

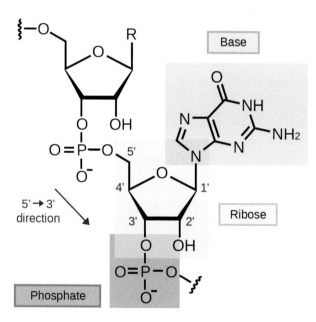

Fig. 5.6 RNA molecule polymerization. RNA molecules bind with each other through the 5′ carbon and an oxygen of the phosphate group. This applies directionality to the molecule which is typically read in a 5′→3′ direction

**Fig. 5.7 Example struc-
ture of an amino acid.**
Structure of a representa-
tive amino acid. All amino
acids contain an amino
group and a carboxyl group
joined to a central carbon
atom. The side group varies
between amino acids and
it is this that gives rise to
their unique chemical and
physical properties

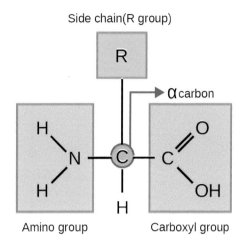

Amino Acids

The building blocks of proteins are amino acids. All amino acids share a common
structure, consisting of an amino group and a carboxyl group connected by an alpha
carbon (Fig. 5.7). A side chain, also known as the R group, differs between the
amino acids with its physical and chemical properties determining the functional
role of the amino acid in the polypeptide. Because of the unique characteristics of
the different side groups, an amino acid can be hydrophobic, hydrophilic, acidic
or basic. There are 20 different amino acids in our proteins with nine described
as essential, meaning they cannot be made in the cells of our body and must be
isolated from our diet.

A polymer of amino acids is called a polypeptide with amino acids joined by
peptide bonds.

Peptide Bonds

A peptide bond (amide bond) is a covalent chemical bond formed between two
amino acid molecules. Amino acids are connected by a dehydration reaction,
marked by the removal of water. The resulting covalent bond is called a peptide bond. A polypeptide, regardless of length, has a single amino acid end (N-
terminus) and a single carboxyl end (C-terminus) (Fig. 5.8).

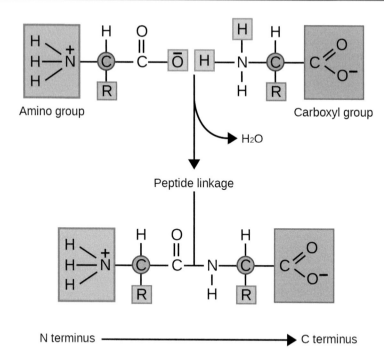

Amino group   Carboxyl group

Peptide linkage

N terminus ———————————————————▶ C terminus

Fig. 5.8   Peptide bond linkage of amino acids. Amino acids are bound by peptide bonds to form polypeptides. The linkage occurs between the amino and carboxyl group releasing H_2O in the process

Proteins

The word protein comes from the Greek word "proteios", which means first or primary. Proteins, the building blocks of life, are synthesized in all forms of living cells. Humans have tens of thousands of types of proteins, which are all constructed from a set of 20 amino acids.

Multiple amino acids connected by peptide bonds form a polypeptide. However, it is important to note that the term polypeptide is not synonymous with protein. A functional protein does not need to be formed just a single polypeptide chain but can actually be made of multiple polypeptides precisely folded into a unique molecular shape. This specific protein structure determines its final function.

Protein Structure

Proteins have at least three structures: primary, secondary, and tertiary, with some having an additional quaternary structure (Fig. 5.9).

The primary structure of a protein is simply its amino acid sequence in a linear form. The secondary structure consists of the coils (alpha-helix) and folds (beta-sheet) that result from the hydrogen bonds which form between repeating constituents of the polypeptide chains. The tertiary structure is the overall shape of the polypeptide resulting from all the interactions between the side chains of various amino acids. A quaternary structure arises when a protein consists of two or more polypeptide chains.

Protein Synthesis

By understanding the basic structure of RNA and proteins we can then understand the process of protein synthesis. Translation refers to the synthesis of a polypeptide using information from an mRNA molecule. Translation takes place in complex molecular machines known as ribosomes which are typically found on the surface of the rough endoplasmic reticulum.

The ribosome assembles around the mRNA and is read in the $5' \rightarrow 3'$ direction. The mRNA sequence is read in groups of three nucleotides referred to as a codon, for example, UGG or CAC. Each codon is associated with a different amino acid and this is why protein synthesis is referred to as translation, the ribosome is translating the mRNA.

The translation process starts with a specific codon known as the start codon, AUG, with this triplet encoding the amino acid methionine. The start codon sets the reading frame for the rest of the mRNA. The following codons are each translated into a specific amino acid, which is added to the C-terminus of the newly formed peptide chain. Translation occurs in three steps: initiation, elongation, and termination. Termination occurs when the ribosome reaches a stop codon, for example, UAG (Fig. 5.10).

Ribosomes

The ribosome is the cellular structure in which translation occurs, with ribosomes located on the surface of the rough endoplasmic reticulum. Ribosomes have a large

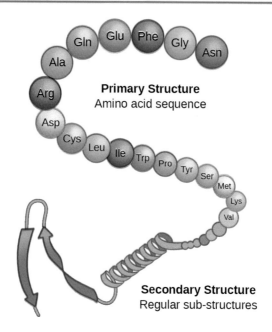

Primary Structure
Amino acid sequence

Secondary Structure
Regular sub-structures

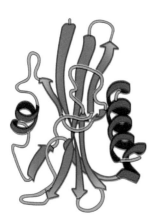

Tertiary Structure
Three-dimensional structure

Quaternary Structure
complex of protein molecules

Fig. 5.9 Schematic view of the different protein structures. The various structures formed by amino acids and proteins are given specific terms. The primary structure refers to the linear amino acid sequence. The secondary structure refers to the formation of regular substructures such as alpha helices or beta sheets. The tertiary structure describes the 3D structure of the protein, accounting for the way the substructures interact with each other. Finally, if a protein is comprised of multiple polypeptides, their interaction is described in the quaternary structure

and small subunit, each constructed from ribosomal RNA (rRNA) and several proteins. Protein synthesis occurs as a result of collaboration between tRNA, mRNA and the ribosome. The mRNA enters the ribosome and codons are read at three sites within the ribosome A (amino acid), P (polypeptide) and E (exit).

When a start codon (AUG) is detected the first tRNA with a complementary anticodon (UAC) binds to the mRNA at the A site. This tRNA is bound to a methionine amino acid which will form the first part of the amino acid chain.

The mRNA moves through the ribosome and this first codon moves to the P site where the methionine amino acid is detached forming the start of the amino acid chain. The next codon enters the A site with its amino acid attached.

The first codon then moves to the E site where the tRNA (without any amino acid bound) is released, this tRNA then binds with a new methionine so it can be reused multiple times. The second codon enters the P site, and the attached amino acid is bound to the already present methionine via a peptide bond.

The mRNA molecule continues to move through the ribosome, with tRNA molecules cycling and bringing the required amino acids until a stop codon is reached at which point translation is terminated and the completed polypeptide released.

Genetic Code

The genetic code relates DNA to mRNA and mRNA to the protein. A codon refers to the three letters of mRNA that encode one amino acid. All possible combinations of the four available base pairs give 64 (4^3) different codons. With this many possibilities, the codons can be redundant but not ambiguous. This means one amino acid can be represented by more than one codon, but a codon can only specify one amino acid. This code specifies the amino acid that will be used to make up a protein (Fig. 5.11).

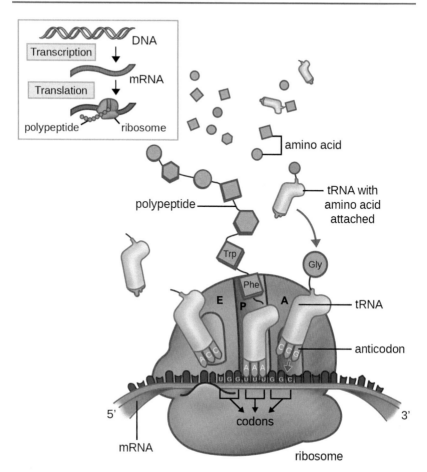

Fig. 5.10 mRNA translation occurring in the ribosome. Ribosomes are responsible for translating mRNA into an amino acid sequence. Composed of a large and small subunit, ribosomes read the mRNA sequence and recruit complementary tRNAs which carry amino acids. mRNA enters the ribosome and codons are read at three sites within the ribosome *A* (amino acid), *P* (polypeptide) and *E* (exit). tRNAs carrying amino acids are recruited to the *A* site. The ribosome moves along the mRNA and this tRNA releases its bound amino acid at the *P* site, which is then attached to the forming polypeptide. Finally, the tRNA is released from the *E* site where it can bind with a new amino-acid and be used again

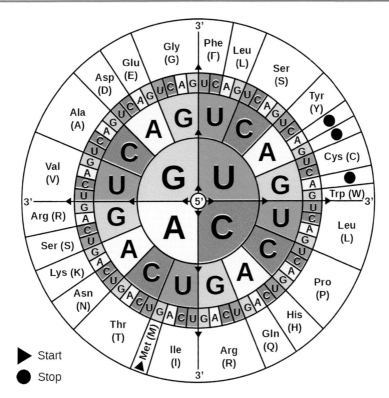

Fig. 5.11 Codon to amino acid translation wheel. The codons in this Fig. are mRNA codons, meaning they are complementary and antiparallel to the base sequence in the DNA strand. To read the circular codon table, start from the letter in the center and follow the codon, for example, UUC is Phe (F). There are four special codons, AUG or the start codon which must be used to start translation, and UAA, UAG and UGA which are the stop codons which terminate translation

Chronic Kidney Disease

Our understanding of protein synthesis has allowed scientists to develop several novel therapies for previously untreatable disorders, such as anemia associated with chronic kidney disease. Chronic kidney disease is a general term for a heterogeneous disorder affecting the function and structure of the kidney. It is difficult to determine the exact cause of chronic kidney disease; however, chronic kidney disease is often associated with old age, hypertension, diabetes, obesity and cardiovascular disease. Most chronic kidney disease patients eventually become anemic, and the severity of anemia increases as the functionality of the kidney decreases. The hormone erythropoietin (EPO) was identified as a potential therapy for people with anemia as it increases the oxygen-carrying capacity of their blood by stimulating the formation of new blood cells.

Erythropoietin

EPO is a glycoprotein hormone produced in the kidney which promotes red blood cell production. A single gene located in chromosome 7 regulates the production of EPO, which primarily takes place in the kidneys. Approximately 10% of EPO is synthesized in the liver and other extra-renal organs.

Structure of EPO

When produced in the cell EPO is a protein 193-amino acids in length, without biological activity. Shortly before secretion, the N-terminus leader sequence of 27 residues is cleaved, leaving only 166 amino acids. When EPO enters the bloodstream, the amino acid arginine is cleaved, leaving a final sequence of 165 amino acids.

This functional EPO contains two disulfide bridges formed by the amino acid cysteine at position 7 and 161, as well as 29 and 33.

The molecular weight of a complete precursor EPO peptide is 18,000 Dalton. After undergoing post-translational modification; glycosylation, whereby a carbohydrate is attached to the protein, the molecular weight of EPO increases to 30,000 Dalton. 40% of the molecular weight of EPO is composed of three N-linked carbohydrate structures, located at Asp-24, 38, and 83, and one O-linked carbohydrate structure, located at Ser-126.

Function of EPO

The EPO molecules that enter the bloodstream are carried to the bone marrow. After EPO reaches the erythrocyte progenitor cells, it triggers the production of erythrocytes by stimulating the proliferation and differentiation of erythroid precursors in different stages, the first stages including burst-forming units (BFU-E) and the second stage including colony forming units (CFU-E) (Fig. 5.12).

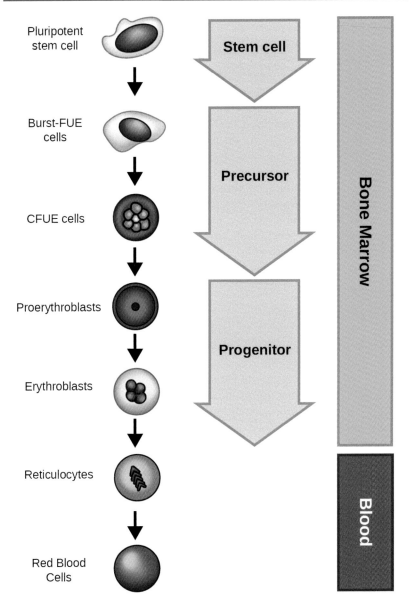

Fig. 5.12 Formation of red blood cells. Red blood cells are ultimately formed from pluripotent stem cells in the bone marrow. However, several distinct phases of division and differentiation occur first meaning that each stem cell gives rise to a huge number of individual red blood cells

Hypoxia triggers the production of EPO. Hypoxia is a condition in which the body, or a region of the body, is deprived of adequate oxygen supply.

Recombinant EPO

Human EPO was first purified in 1977 from 2,500 liters of urine from an anemic patient. Following this, in 1985 the human EPO cDNA sequence was mapped. Today, with advances in molecular genetics engineering, it is possible to synthesize recombinant human EPO in various cell types.

Recombinant human EPO synthesized in *E. coli* is unstable at high temperatures and tends to aggregate as bacteria are unable to glycosylate proteins. The mammalian cell system is the only expression system that can perfectly duplicate the complex structure of biological active EPO that comprises two disulfide bridges and four glycosylation sites. The recombinant human EPO synthesized in mammalian cells is more stable due to its 40% carbohydrate component.

Applications of rhEPO

The US Food and Drug Administration (USFDA) approved recombinant human EPO (rhEPO) in 1989 for the treatment of anemia in association with chronic kidney disease. This is because rhEPO promotes red blood cell production. Even though rhEPO is used for therapeutic purposes, it can also be abused, such as in the case of blood doping in sporting events.

EPO can stimulate red blood cell production and thus increase oxygen delivery to tissue. This is because hemoglobin resides in red blood cells, its primary role being to bind oxygen and transport it to areas where oxygen is needed. The level of oxygen plays a critical role in athletic performance, where higher oxygen levels lead to a higher aerobic metabolism that can generate higher energy (ATP), putting the athlete who is doped at a competitive advantage.

The International Olympic Committee prohibits the administration of EPO, and its use is classified as doping.

Detection of rhEPO

Detection of rhEPO is conducted using a mass spectrometer. The mass-to-charge ratio (m/z) is roughly the same as the protein mass in Daltons. We use a urine sample for substance detection. In most cases, it is better to use a urine sample instead of blood to test for prohibited substances. This is because urine collection is non-invasive and yields a large sample volume with a higher drug concentration than blood. Also, urine contains fewer cells and proteins that would complicate extraction.

To detect prohibited substances in urine, one has to first make a standard spectrum of uncontaminated urine (negative standard), as well as of the prohibited substance (positive standard). Once these two standards are generated, one can compare them with a sample spectrum. If the sample spectrum exhibits the same peaks that the positive standard has, one can conclude that the sample contains the prohibited substance. Mass spectrometry can also detect glycosylation, as the data from the mass spectrometer is processed and displayed in peaks. Usually, glycosylation peaks are marked with an asterisk or a glycan diagram.

Mass Spectrometry

In the protein synthesis simulation you will use mass spectrometry to detect rhEPO in urine samples from cyclists competing in a race.

A mass spectrometer is an analytical tool used to measure the molecular mass of a sample. The three fundamental parts of a mass spectrometer are the ionization source, analyzer, and the detector. A mass spectrometer functions by:

- Producing ions from the sample in the ionization source.
- Separating these ions according to their mass-to-charge ratio in the mass analyzer.
- Fragmenting the selected ions and analyzes the fragments in a second analyzer.
- Detecting the ions emerging from the last analyzer and measures their abundance with the detector that converts the ions into electrical signals.
- Processing the signals from the detector that are transmitted to the computer and controls the instrument through feedback.

Matrix-assisted Laser Desorption Ionization (MALDI)

There are a variety of ionization methods available, however, the most commonly used methods are: Electrospray ionization (ESI) and MALDI. In this lab we use MALDI.

There are two major steps involved in MALDI, which uses a UV-laser to ionize a sample. Firstly, the peptide to be analyzed is dissolved in a matrix, which is a solvent containing small organic molecules. These molecules have high absorption at the laser's wavelength. The mixture is then left to dry prior to analysis. This results in the formation of a crystal matrix that contains the peptide of interest. Secondly, the matrix molecule is excited to a higher energy state when it encounters the UV-laser. This eventually leads to the formation of the ion of interest. The ionized molecule enters the mass analyzer and yields the mass spectrum.

Time of Flight (TOF)

The MALDI ionization method is coupled with a mass analyzer called time-of-flight (TOF). In TOF, the time ions take to reach the detector is measured. The velocity is determined by the mass-to-charge ratio (m/z). Smaller ions will reach the detector before large ones due to the fact they have less mass. The charge of the ions also determines the velocity. Ions with two or more positive charge will move faster than ions with only one positive charge. Thus, those ions that are both smaller and positively charged will move faster.

The output of the detector is the mass spectrum, displayed in a "stick diagram". This shows the relative current produced by ions of varying mass/charge ratios.

5.3   Let's Get Started

Wow, what a journey! In the final step of this Basic Biology textbook you will be able to follow the journey of EPO from DNA to protein and learn how to produce it in the lab. While rhEPO has great health benefits for some, others use it for more nefarious purposes. Will you be able to use your knowledge of mass spectrometry to catch the doping cheat in the protein synthesis lab?

Techniques Used in the Lab
- Protein synthesis
- Mass spectrometry

Learning Objectives
At the end of this simulation, you will be able to...

- Describe the process of protein synthesis all the way from transcription of DNA to mRNA, to translation of mRNA into a protein followed by final alterations to the protein structure
- Classify the primary, secondary, tertiary and quaternary structures of protein
- Design an experiment to produce a recombinant protein using bacterial or animal cells
- Utilize mass spectrometry to identify recombinant proteins of interest

ACCESS THE VIRTUAL LAB SIMULATION HERE www.labster.com/springer **BY USING THE UNIQUE CODE AT THE END OF THE PRINTED BOOK. IF YOU USE THE E-BOOK YOU CAN PURCHASE ACCESS TO THE SIMULATIONS THROUGH THE SAME LINK.**

Further Reading

Alberts B et al (2015) The Molecular Biology of the Cell, 6th edn. Garland Science, Abingdon

Bento, R.M.d.A., Damasceno L.M.P., and Aquino Neto F.R.d. *Recombinant human erythropoietin in sports: a review.* Revista Brasileira de Medicina do Esporte, 2003. 9: p. 181–190.

Delanghe JR, Bollen M, Beullens M (2008) Testing for recombinant erythropoietin. Am J Hematol 83(3):237–241

Hatton CK (2007) Beyond Sports-Doping Headlines: The Science of Laboratory Tests for Performance-Enhancing Drugs. Pediatr Clin North Am 54(4):713–733

Jelkmann W (1992) Erythropoietin: structure, control of production, and function. Physiol Rev 72(2):449–489

Nelson DL, Cox MM, Lehninger AL (2013) Lehninger Principles of Biochemistry, 6th edn. Freeman W.H, New York

OpenStax, Biology. OpenStax CNX. Jun 1, 2018, http://cnx.org/contents/185cbf87-c72e-48f5-b51e-f14f21b5eabd@11.2

Urey LA et al (2014) Campbell Biology, 10th edn. Pearson, Boston

Activate your free access to the LABSTER simulations

You can access the virtual lab simulations included in this book at
www.labster.com/springer.

The following code gives you free access to the simulations for the duration of one
semester (six months).
Please be aware that the six month period starts once you sign in for the first time.

your personal code: 6Z26LMNUK5

If you have problems using the voucher, you can contact us at
customerservice@springer.com.

Printed by Printforce, the Netherlands